全国餐饮职业教育教学指导委员会重点课题“基于烹饪专业人才培养目标的中高职课程体系与教材开发研究”成果系列教材
餐饮职业教育创新技能型人才培养新形态一体化系列教材

总主编 ◎杨铭铎

烹饪英语基础

主　编　张　华　刘　明　谢碧霞
副主编　韩　燕　李　徽　董　立　覃　怀
编　者（按姓氏笔画排序）
刘　明　李　徽　张　华　张　睿
陈恩程　董　立　韩　燕　覃　怀
谢碧霞　褚晓霞

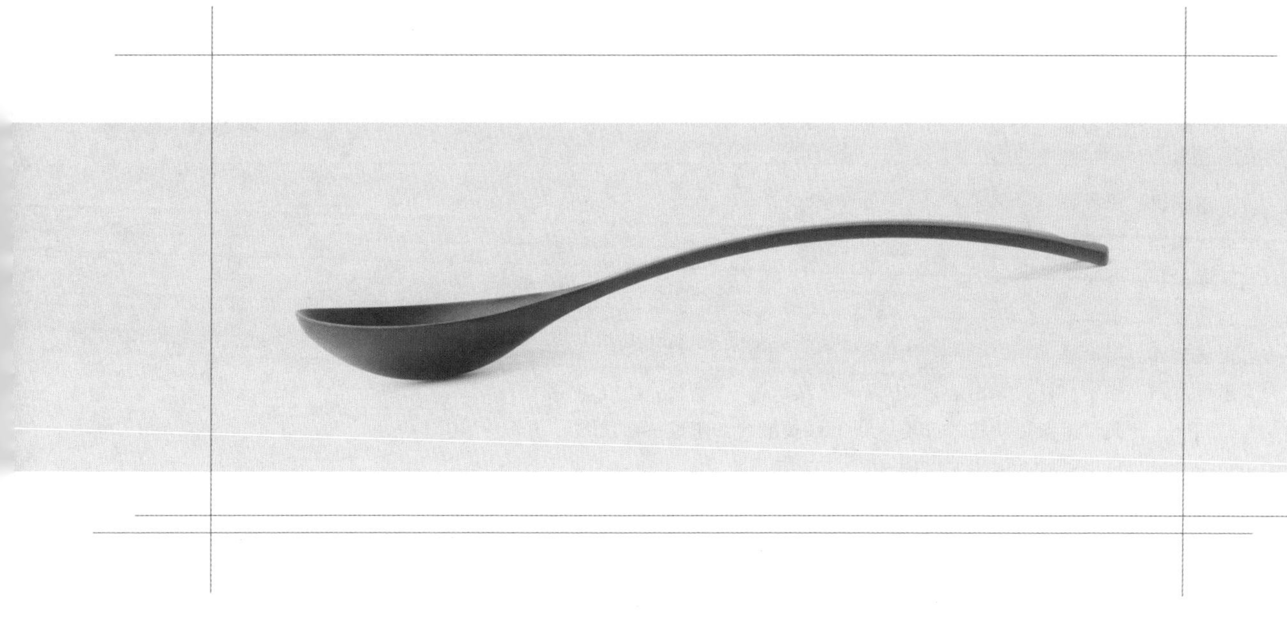

華中科技大學出版社
http://www.hustp.com
中国·武汉

内 容 简 介

本教材为全国餐饮职业教育教学指导委员会重点课题"基于烹饪专业人才培养目标的中高职课程体系与教材开发研究"成果系列教材和餐饮职业教育创新技能型人才培养新形态一体化系列教材。

本教材共7个单元26课，内容包括厨房、原料加工、中国菜品制作、国外菜品制作、中西式点心、菜品装饰、中西餐服务等，文末还附有词汇表，供学生学习时使用。本教材用生动形象、通俗易懂的图片、对话、故事等介绍烹饪英语基础知识，使教材更具可读性，同时配备英文听力材料、教学课件等丰富的数字教学资源。

本教材适用于职业院校烹饪(餐饮)专业，也可作为餐饮行业从业人员的烹饪英语入门用书。

图书在版编目(CIP)数据

烹饪英语基础/张华，刘明，谢碧霞主编. —武汉：华中科技大学出版社，2020.6(2024.1重印)
ISBN 978-7-5680-6200-8

Ⅰ. ①烹… Ⅱ. ①张… ②刘… ③谢… Ⅲ. ①烹饪-英语-职业教育-教材 Ⅳ. ①TS972.1

中国版本图书馆CIP数据核字(2020)第087224号

烹饪英语基础
Pengren Yingyu Jichu

张 华 刘 明 谢碧霞 主编

策划编辑：汪飒婷
责任编辑：曾奇峰
封面设计：廖亚萍
责任校对：李 琴
责任监印：周治超
出版发行：华中科技大学出版社(中国·武汉) 电话：(027)81321913
武汉市东湖新技术开发区华工科技园 邮编：430223
录 排：华中科技大学惠友文印中心
印 刷：湖北新华印务有限公司
开 本：889mm×1194mm 1/16
印 张：7.25
字 数：210千字
版 次：2024年1月第1版第4次印刷
定 价：39.00元

全国餐饮职业教育教学指导委员会重点课题

“基于烹饪专业人才培养目标的中高职课程体系与教材开发研究”成果系列教材

餐饮职业教育创新技能型人才培养新形态一体化系列教材

丛书编审委员会

委员（按姓氏笔画排序）

网络增值服务

使用说明

欢迎使用华中科技大学出版社医学资源网

教师使用流程

（1）登录网址：http://yixue.hustp.com（注册时请选择教师用户）

注册 → 登录 → 完善个人信息 → 等待审核

（2）审核通过后，您可以在网站使用以下功能：

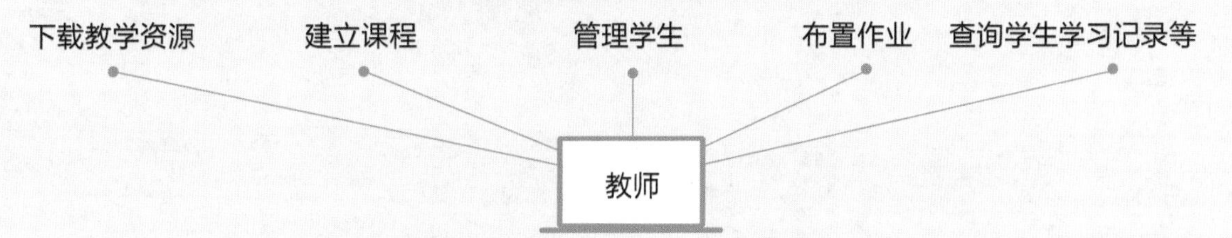

学员使用流程

（建议学员在PC端完成注册、登录、完善个人信息的操作。）

（1）PC端学员操作步骤

①登录网址：http://yixue.hustp.com（注册时请选择普通用户）

注册 → 登录 → 完善个人信息

②查看课程资源：（如有学习码，请在"个人中心—学习码验证"中先通过验证，再进行操作。）

（2）手机端扫码操作步骤

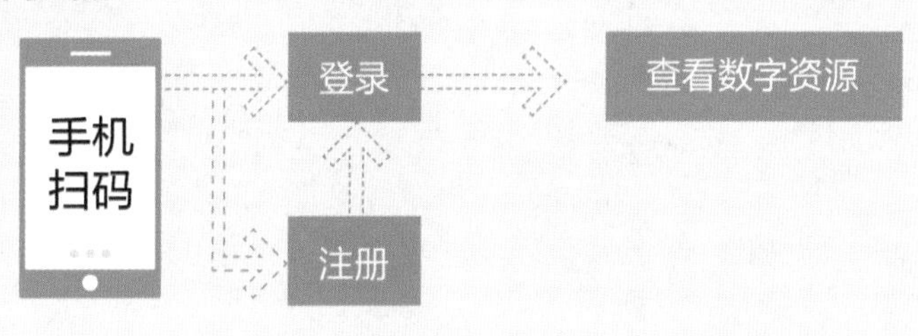

开展餐饮教学研究　加快餐饮人才培养

餐饮业是第三产业重要组成部分，改革开放40多年来，随着人们生活水平的提高，作为传统服务性行业，餐饮业对刺激消费需求、推动经济增长发挥了重要作用，在扩大内需、繁荣市场、吸纳就业和提高人民生活质量等方面都做出了积极贡献。就经济贡献而言，2018年，全国餐饮收入42716亿元，首次超过4万亿元，同比增长9.5%，餐饮市场增幅高于社会消费品零售总额增幅0.5个百分点；全国餐饮收入占社会消费品零售总额的比重持续上升，由上年的10.8%增至11.2%；对社会消费品零售总额增长贡献率为20.9%，比上年大幅上涨9.6个百分点；强劲拉动社会消费品零售总额增长了1.9个百分点。中国共产党第十九次全国代表大会(简称党的十九大)吹响了全面建成小康社会的号角，作为人民基本需求的饮食生活，餐饮业的发展好坏，不仅关系到能否在扩内需、促消费、稳增长、惠民生方面发挥市场主体的重要作用，而且关系到能否满足人民对美好生活的向往、实现全面建成小康社会的目标。

一个产业的发展，离不开人才支撑。科教兴国、人才强国是我国发展的关键战略。餐饮业的发展同样需要科教兴业、人才强业。经过60多年特别是改革开放40多年来的大发展，目前烹饪教育在办学层次上形成了中职、高职、本科、硕士、博士五个办学层次；在办学类型上形成了烹饪职业技术教育、烹饪职业技术师范教育、烹饪学科教育三个办学类型；在学校设置上形成了中等职业学校、高等职业学校、高等师范院校、普通高等学校的办学格局。

我从全聚德董事长的岗位到担任中国烹饪协会会长、全国餐饮职业教育教学指导委员会主任委员后，更加关注烹饪教育。在到烹饪院校考察时发现，中职、高职、本科师范专业都开设了烹饪技术课，然而在烹饪教育内容上没有明显区别，层次界限模糊，中职、高职、本科烹饪课程设置重复，拉不开档次。各层次烹饪院校人才培养目标到底有哪些区别？在一次全国餐饮职业教育教学指导委员会和中国烹饪协会餐饮教育委员会的会议上，我向在我国从事餐饮烹饪教育时间很久的资深烹饪教育专家杨铭铎教授提出了这一问题。为此，杨铭铎教授研究之后写出了《不同层次烹饪专业培养目标分析》《我国现代烹饪教育体系的构建》，这两篇论文回答了我的问题。这两篇论文分别刊登在《美食研究》和《中国职业技术教育》上，并收录在中国烹饪协会主编的《中国餐饮产业发展报告》之中。我欣喜地看到，杨铭铎教授从烹饪专业属性、学科建设、课程结构、中高职衔接、课程体系、课程开发、校企合作、教师队伍建设等方面进行研究并提出了建设性意见，对烹饪教育发展具有重要指导意义。

杨铭铎教授不仅在理论上探讨烹饪教育问题，而且在实践上积极探索。2018年在全国餐饮职业教育教学指导委员会立项重点课题“基于烹饪专业人才培养目标的中高职课程体系与教材开发研究”(CYHZWZD201810)。该课题以培养目标为切入点，明晰烹饪专业人才培养规格；以职业技能为结合点，确保烹饪人才与社会职业有效对接；以课程体系为关键点，通过课程结构与课程标准精准实现培养目标；以教材开发为落脚点，开发教学过程与生产过程对接的、中高职衔接的两套烹饪专业课程系列教材。这一课题的创新点在于：研究与编写相结合，中职与高职相同步，学生用教材与教师用参考书相联系，资深餐饮专家领衔任总主编与全国排名前列的大学出版社相协作，编写出的中职、高职系列烹饪专业教材，解决了烹饪专业文化基础课程与职业技能课程脱节、专业理论课程设置重复、烹饪技能课交叉、职业技能倒挂、教材内容拉不开层次等问题，是国务院《国家职业教育改革实施方案》提出的完善教育教学相关标准中的持续更新并推进专业教学标准、课程标准建设和在职业院校落地实施这一要求在烹饪职业教育专业的具体举措。基于此，我代表中国烹饪协会、全国餐饮职业教育教学指导委员会向全国烹饪院校和餐饮行业推荐这两套烹饪专业教材。

习近平总书记在党的十九大报告中将“两个一百年”奋斗目标调整表述为：到建党一百年时，全面建成小康社会；到新中国成立一百年时，全面建成社会主义现代化强国。经济社会的发展，必然带来餐饮业的繁荣，迫切需要培养更多更优的餐饮烹饪人才，要求餐饮烹饪教育工作者提出更接地气的教学和科研成果。杨铭铎教授的研究成果，为中国烹饪技术教育研究开了个好头。让我们餐饮烹饪教育工作者与餐饮企业家携起手来，为培养千千万万优秀的烹饪人才、推动餐饮业又好又快的发展，为把我国建成富强、民主、文明、和谐、美丽的社会主义现代化强国增添力量。

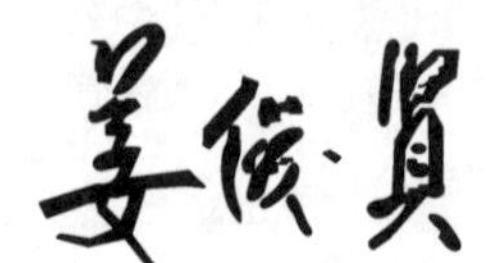

全国餐饮职业教育教学指导委员会主任委员

中国烹饪协会会长

《国家中长期教育改革和发展规划纲要(2010—2020年)》及《国务院办公厅关于深化产教融合的若干意见(国办发〔2017〕95号)》等文件指出:职业教育到2020年要形成适应经济发展方式的转变和产业结构调整的要求,体现终身教育理念,中等和高等职业教育协调发展的现代教育体系,满足经济社会对高素质劳动者和技能型人才的需要。2019年1月,国务院印发的《国家职业教育改革实施方案》中更是明确提出了提高中等职业教育发展水平、推进高等职业教育高质量发展的要求及完善高层次应用型人才培养体系的要求;为了适应"互联网+职业教育"发展需求,运用现代信息技术改进教学方式方法,对教学教材的信息化建设,应配套开发信息化资源。

随着社会经济的迅速发展和国际化交流的逐渐深入,烹饪行业面临新的挑战和机遇,这就对新时代烹饪职业教育提出了新的要求。为了促进教育链、人才链与产业链、创新链有机衔接,加强技术技能积累,以增强学生核心素养、技术技能水平和可持续发展能力为重点,对接最新行业、职业标准和岗位规范,优化专业课程结构,适应信息技术发展和产业升级情况,更新教学内容,在基于全国餐饮职业教育教学指导委员会2018年度重点课题"基于烹饪专业人才培养目标的中高职课程体系与教材开发研究"(CYHZWZD201810)的基础上,华中科技大学出版社在全国餐饮职业教育教学指导委员会副主任委员杨铭铎教授的指导下,在认真、广泛调研和专家推荐的基础上,组织了全国90余所烹饪专业院校及单位,遴选了近300位经验丰富的教师和优秀行业、企业人才,共同编写了本套餐饮职业教育创新技能型人才培养新形态一体化系列教材、全国餐饮职业教育教学指导委员会重点课题"基于烹饪专业人才培养目标的中高职课程体系与教材开发研究"成果系列教材。

本套教材力争契合烹饪专业人才培养的灵活性、适应性和针对性,符合岗位对烹饪专业人才知识、技能、能力和素质的需求。本套教材有以下编写特点:

1. 权威指导,基于科研　本套教材以全国餐饮职业教育教学指导委员会的重点课题为基础,由国内餐饮职业教育教学和实践经验丰富的专家指导,将研究成果适度、合理落脚于教材中。

2. 理实一体,强化技能　遵循以工作过程为导向的原则,明确工作任务,并在此基础上将与技能和工作任务集成的理论知识加以融合,使得学生在实际工作环境中,将知识和技能协调配合。

3. 贴近岗位,注重实践　按照现代烹饪岗位的能力要求,对接现代烹饪行业和企业的职业技能标准，将学历证书和若干职业技能等级证书（"1＋X"证书）内容相结合，融入新技术、新工艺、新规范、新要求，培养职业素养、专业知识和职业技能，提高学生应对实际工作的能力。

4. 编排新颖，版式灵活　注重教材表现形式的新颖性，文字叙述符合行业习惯，表达力求通俗、易懂，版面编排力求图文并茂、版式灵活，以激发学生的学习兴趣。

5. 纸质数字，融合发展　在新形势媒体融合发展的背景下，将传统纸质教材和我社数字资源平台融合，开发信息化资源，打造成一套纸数融合的新形态一体化教材。

本系列教材得到了全国餐饮职业教育教学指导委员会和各院校、企业的大力支持和高度关注，它将为新时期餐饮职业教育做出应有的贡献，具有推动烹饪职业教育教学改革的实践价值。我们衷心希望本套教材能在相关课程的教学中发挥积极作用，并得到广大读者的青睐。我们也相信本套教材在使用过程中，通过教学实践的检验和实际问题的解决，能不断得到改进、完善和提高。

随着我国餐饮业的快速发展及其逐步与国际标准接轨，餐饮企业对从业人员的英语综合素质提出了更高的要求。在全国餐饮职业教育教学指导委员会2018年重点课题“基于烹饪专业人才培养目标的中高职课程体系与教材开发研究”立项之时，我们将专业、课程、教材紧密联系，整合优势资源，组织各地中职院校具有丰富经验的一线教师编写了本教材。本教材共7个单元26课，涉及厨房、原料加工、中国菜品制作、国外菜品制作、中西式点心、菜品装饰、中西餐服务等内容。

本教材在整体设计上与传统教材的不同之处有以下两点。一是本教材不以学科逻辑划分篇章，而是基于厨房、原料、菜品和点心制作、菜品装饰、中西餐服务的流程来划分全书的结构，用生动形象、通俗易懂的图片、对话等使学生加深对烹饪行业的理解。二是本教材将烹饪各方面知识内容协调处理，以适当的形式呈现出来。尤其在呈现方式上，基于职业教育层次的学生对知识性较强的内容阅读起来会较困难，因而本教材尽量将其转变为图表、图片、讲故事等形式，增加对学生的吸引力，减少学生对知识学习的畏惧感，以便学生更好地理解和接受。

本教材第1、2、23、24课由海南省技师学院韩燕老师编写，第3、15、19课由重庆现代职业技师学院张华老师编写，第4、5、6、7课由西安商贸旅游技师学院李徽老师编写，第8、13课由珠海市第一中等职业学校刘明老师编写，第9、25、26课由重庆现代职业技师学院董立老师编写，第10、14课由佛山市顺德区梁銶琚职业技术学校谢碧霞老师编写，第11、12课由太原市财政金融学校张睿老师编写，第16、17、18、20课由珠海市第一中等职业学校陈恩程老师编写，第21、22课由西安商贸旅游技师学院褚晓霞老师编写。云南能源职业技术学院覃怀老师负责本教材部分数字资源的开发和音频录制工作。建议单元1、2、4、6、7每课两学时，单元3、5每课4学时，共72学时。

编者在本教材的编写过程中参阅了大量国内已出版的有关资料，限于篇幅，我们没有一一注明出处，主要参考书目附于书末，希望以此表达对这些编著者的诚挚谢意。同时，本教材的编写和出版得到了华中科技大学出版社的大力支持和帮助，在此也向他们致以衷心的感谢。

由于编者水平有限，书中难免出现错误和不足，请广大读者批评指正。

编者

目录

Unit 1

Kitchen

Lesson 1 Professional Chef

扫码看课件

Goal

You will be able to:

1. Get acquaintance with chef duties.
2. Greet to others.

Warming up

❶ Read and match

Prep cook	Pastry chef	Cold dishes chef
Hot dishes chef	Larder chef	Executive chef

1. ________________

2. ________________

3. ________________

4. ________________

5. ________________

6. ________________

❷ Learn and say

1. He is responsible for baking pastries.
He is a ________________.
2. He is in charge of fried hot dishes.
He is a ________________.
3. He works under a chef to learn to work.
He is a ________________.
4. He is responsible for preparing cold dishes.
He is a ________________.
5. He is in charge of grill meat.

Note

He is a ________________.

6. He oversees all kitchen staff.

He is a ________________.

Let's learn

Section A

外教有声

① Listen to the dialogue and repeat

Jim:Good morning,Lily.

Lily:Good morning,Jim. Welcome to join us. Let me show you to your trainer.

Jim:OK. Thank you very much!

(Lily took Jim to see Jack.)

Lily:Hi,Jack! This is Jim. He is our new cook.

Jack:How do you do,Jim?

Jim:How do you do,Jack?

Lily:Jim,Jack is your boss. He will train you in the following 3 months. Please learn from him by heart. And I believe you'll make a great progress soon.

Jim:I will. Thank you very much!

Lily:You're welcome.

② Complete the dialogue

A:Hi,Jack! __________ is Jim.

B:________________, Jim? I am Jack.

C:________________, Jack?

③ Learn and say

Jack:There are mainly 6 kinds of chefs in our restaurant.

Jim:What are they?

Jack:Prep cook,pastry chef,cold dishes chef,hot dishes chef,larder chef and executive chef.

Jim:I see.

Jack:Now,I'll show you to our colleagues. This way,please.

A:What does a prep cook do?

B:He ________________.

Section B

① Learn more

Executive chef 行政总厨
↓
Head chef 厨师长
↓
Assistant head chef 厨师长助理
↓
Sauce chef 调味厨师 | Sauté chef 烹炒厨师 | Pastry chef 面点厨师 | Roast chef 烧烤厨师 | Vegetable chef 蔬菜厨师
↓
Assistant chef 厨师助理 | Cook 厨工

❷ **Write down chefs in a restaurant kitchen in this lesson and then translate into Chinese and write down their duties**

English	Chinese	Duties
Executive chef	行政总厨	

Self-check

Words I have learned in this lesson are:

☐ assistant ☐ chef ☐ cook

☐ executive ☐ pastry ☐ progress

☐ roast ☐ sauce ☐ vegetable

I know ________ words.

Phrases and expressions I have learned in this lesson are:

☐ fried hot dishes ☐ baking pastries

☐ works under a chef ☐ prepare cold dishes

☐ grill meat ☐ oversees all kitchen staff

I know ________ phrases and expressions.

I can:

☐ greet to others

☐ get acquaintance with chef duties

Lesson 2 Kitchen Equipment and Tools

Goal

You will be able to:

1. Identify the kitchen equipment.
2. Identify the kitchen tools.
3. Know the usage of the kitchen equipment and tools.

Warming up

❶ Read and match kitchen equipment

Refrigerator	Microwave oven	Gas stove
Disinfection cabinet	Meat slicer	Smoking machine
Blender	Barbecue grill	Automatic dishwasher

1. ____________

2. ____________

3. ____________

4. ____________

5. ____________

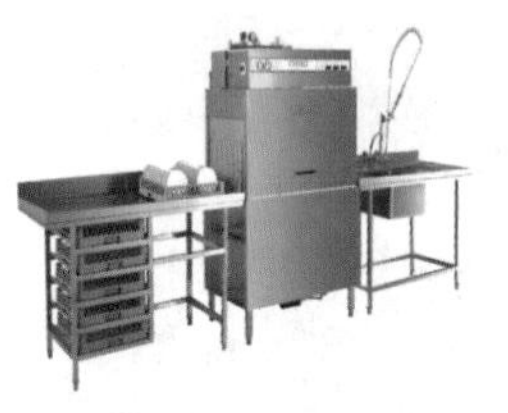

6. ____________

7. ____________

8. ____________

9. ____________

❷ Learn and practice more

A: What equipment do you have?

B: I have ________.

A: What is it used for?

B: It is used for ________.

Let's learn

外教有声

Note

Section A

❶ Listen to the dialogue and repeat 

A: I want to reheat these fried greens. What should I do?

B: You can reheat it in a special microwave oven bowl. Turn on the microwave oven, low heat for 1 minute.

A: Oh, it's very simple. Any other ways should I take?

B: You can pour it into a pot and then switch on the gas stove, high heat for 2 minutes.

A: OK, I got it. Thank you very much!

B: You're welcome.

② Learn and choose

1. Please power on/open the blender.
2. Please switch off/close the gas stove.
3. Please turn on/open the door of refrigerator.
4. Please turn off/close the door of the dishwasher.

Section B

① Read and match kitchen tools

Spoon	Steamer	Chopsticks	Electronic scale	
Knife	Wok	Pan	Rolling pin	Fork

1. ________ 2. ________ 3. ________

4. ________ 5. ________ 6. ________

7. ________ 8. ________ 9. ________

② Fill and read

1. I can use ________ to fry egg.
2. I can use ________ to steam fish.
3. I can use ________ to stew meat.
4. I can use ________ to eat soup.
5. I can use ________ to pick dishes.
6. I can use ________ to fork bread.
7. I can use ________ to weigh food.
8. I can use ________ to cut fruit.
9. I can use ________ to roll paste.

Note

扫码看答案

Self-check

Words I have learned in this lesson are:

□ automatic	□ barbecue	□ blender	□ cabinet
□ chopsticks	□ disinfection	□ dishwasher	□ electronic
□ grill	□ grinder	□ microwave	□ oven
□ refrigerator	□ scale	□ slicer	□ spoon
□ steamer	□ stove		

I know __________ words.

Phrases and expressions I have learned in this lesson are:

□ automatic dishwasher	□ be used for
□ disinfection cabinet	□ gas stove
□ meat slicer	□ turn on/off

I know __________ phrases and expressions.

I can:

□ identify common kitchen equipment

□ learn to use kitchen tools

Unit 2

Cooking Materials

Lesson 3 Vegetables and Fruits

扫码看课件

Goal

You will be able to:

1. Master the expressions of vegetables and fruits.
2. Describe how to make fruit salad.
3. Classify the warm season crops and the cool season crops.

Warming up

1 Read and match

Kiwifruit	Turnip	Lettuce	Celery	Eggplant
Broccoli	Strawberry	Mango	Pineapple	

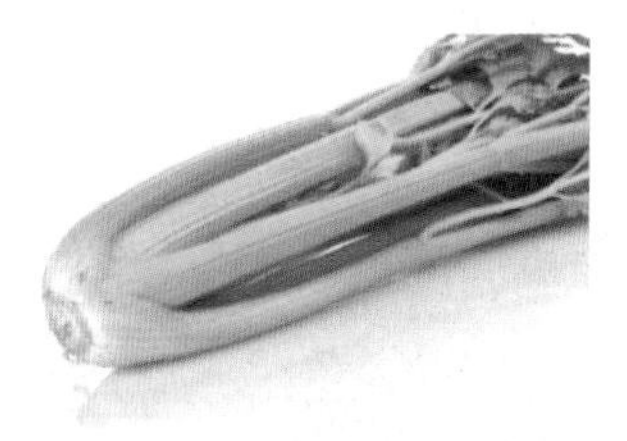

1. ________________

2. ________________

3. ________________

4. ________________

5. ________________

6. ________________

7. ________________

8. ________________

9. ________________

❷ Learn and say

A: What vegetables do you like?

B: I like lettuce and celery.

A: What fruits do you have?

B: I have a pineapple and some strawberries.

❸ Learn and act

A: What vegetables do you like?

B: I like ________.

A: What fruits do you have?

B: I have ________.

Let's learn

Section A

外教有声

❶ Listen to the dialogue and repeat 

Lily: Bill, I am hungry. Let's start to make fruit salad now!

Bill: That sounds good.

Lily: What fruits do we have?

Bill: We have some bananas, apples, kiwifruits and oranges.

Lily: OK, that's enough. First, let's wash and peel them all.

Bill: And then?

Lily: Cut the fruits into different nice shapes. Put them in a bowl.

Bill: Anything else?

Lily: We also need yogurt. Pour it into the bowl. Finally, mix them up.

Bill: The fruit salad looks good. I can't wait to try it.

❷ Fill in the blanks with the verbs you have learned in this dialogue

1.______ the grapes

2.______ the potatoes

3.______ the eggplants

4.______ the oranges into the bag

5.______ the fruits

6.______ the water into the glass

❸ Put the sentences below into the right order

__________ Finally, mix them up.

__________ And pour some yogurt into the bowl.

__________ Then, peel the pineapple and the apple.

__________ First, wash a pineapple and an apple.

__________ Next, cut them into small cubes.

Section B

❶ Read the short passage

Vegetables may be classified as warm season crops and cool season crops. Warm season crops need high temperatures in order to grow well. Some warm season vegetables are cucumbers, pumpkins, string beans and tomatoes. Cool season crops are able to grow when the temperatures are low. They can be planted early in May. Some cool season crops are beets, carrots, lettuces, peas and turnips.

❷ What vegetables can you see in the pictures? Please write them down

1.________________ 2.________________ 3.________________

4.________________ 5.________________ 6.________________

扫码看答案

❸ Classify the vegetables in the above pictures

Warm season crops: ______

Cool season crops: ______

Self-check

Words I have learned in this lesson are:

□ beet	□ broccoli	□ carrot	□ celery
□ cucumber	□ eggplant	□ fruit	□ kiwifruit
□ lettuce	□ mango	□ melon	□ pea
□ peel	□ pineapple	□ pour	□ pumpkin
□ strawberry	□ turnip	□ temperature	

I know ______ words.

Phrases and expressions I have learned in this lesson are:

□ cut... into...

□ mix up

□ make fruit salad

□ peel the banana

□ pour... into...

□ put... into...

□ warm/cool season crops

□ What fruits/vegetables do you have?

I know ______ phrases and expressions.

I can:

□ describe how to make fruit salad

□ classify the warm season crops and the cool season crops

Lesson 4 Poultry

扫码看课件

Goal

You will be able to:

1. Master the names of different meats.
2. Describe some basic cutting skills of meat.
3. Describe some basic cooking skills of meat.

Warming up

❶ Read and match

Goose	Chicken wing	Duck neck
Chicken breast	Duck	Chicken thigh
Goose web	Chicken	Quail

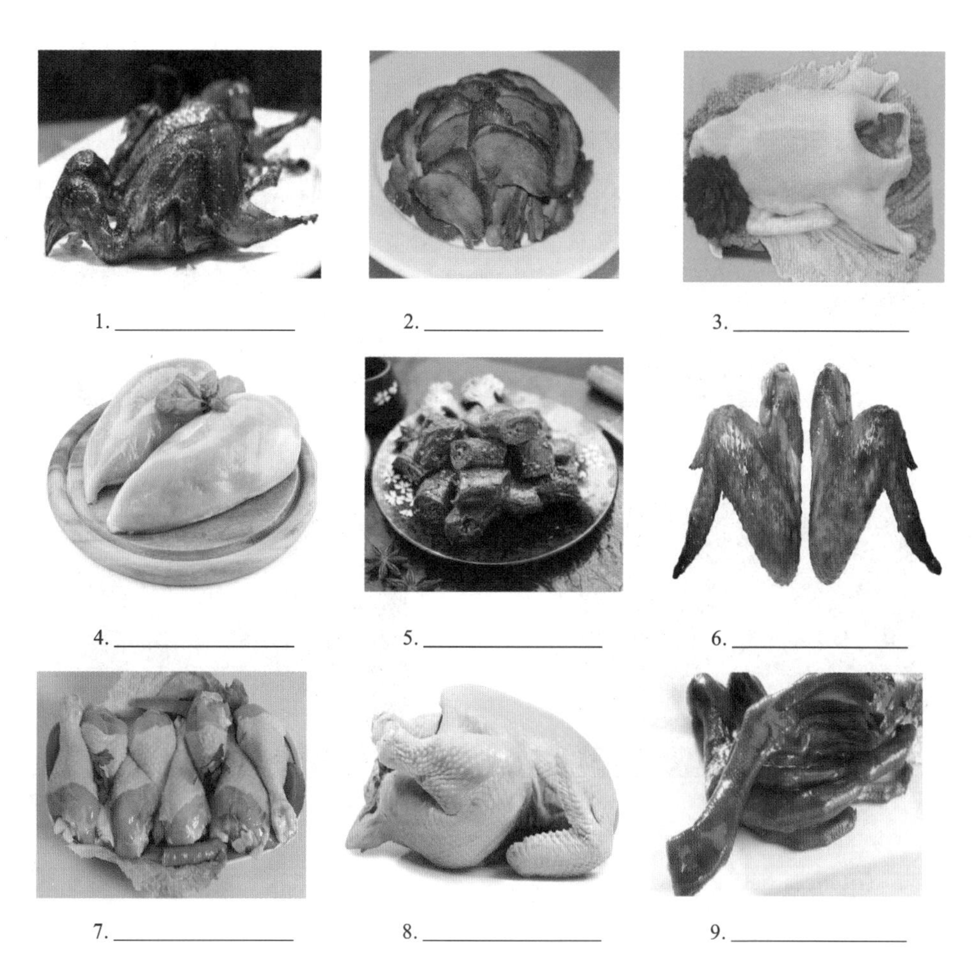

1. ________ 2. ________ 3. ________

4. ________ 5. ________ 6. ________

7. ________ 8. ________ 9. ________

❷ Learn and say

A: Which one do you like better, chicken or duck?

B: I like chicken better.

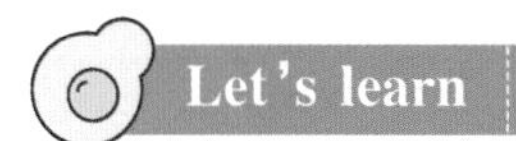

Let's learn

Section A

❶ Listen to the dialogue and repeat 

外教有声

Lynne: Peter, this is my first time to Xi'an, could you please recommend some delicious food to me?

Peter: Sure, Xi'an is famous for its local food. Let me share with you some interesting food today. Have you ever heard of the gourd chicken?

Lynne: Never, what's that?

Peter: People here make the chicken tender and crisp. I'll never forget the fragrance.

Lynne: Wow, cool. I like Chinese food. Would you mind telling me how to cook this dish?

Peter: OK, the preparation procedure is complicated. I'll give you a brief introduction. First, the chicken should be cleaned and boiled in the water to get rid of its blood. Second, steam the chicken with some ingredients such as rice wine, soybean, salt, ginger, scallion, aniseed and cinnamon. Finally, fry it until it turns golden yellow, and then make it a guard-shape when serving.

Lynne: That sounds complex.

Peter: Yes, but it's quite worth your energy.

❷ Fill in the blanks with the verbs you have learned in this dialogue

1. ______ the pumpkin

2. ______ the chicken thigh

3. ______ the egg

4. ______ the dish

❸ Put the sentences below into the right order

__________ Finally, make it a guard-shape when serving.

__________ Third, fry the chicken until it turns golden yellow.

__________ Second, steam the chicken with some ingredients.

__________ First, clean the chicken.

Section B

❶ Read the recipe

Kung pao chicken

Ingredients:

200 grams of chicken breast, 50 grams of fried peanuts, 30 grams of scallion, 20 grams of sliced ginger, 5 grams of red chili, 5 grams of cooking wine, salt, soy sauce, 10 mL starch, an egg white.

Directions:

1. Flatten and dice the chicken breast.
2. Marinate the diced chicken with salt, egg white and starch.
3. Fry the diced chicken until medium-well.
4. Stir-fry the red chili, sliced ginger and scallion, and then add the diced chicken.
5. Put in the fried peanuts, soy sauce and cooking wine.

❷ Complete the cooking process

1. ______________ and ______________ the chicken breast.

2. ________________ the diced chicken with ________________.

3. ________________ the diced chicken until ________________.

4. ________________ the red chili, sliced ginger and scallion, and then ________________ the diced chicken.

5. __________ the __________, soy sauce and cooking wine.

Self-check

Words I have learned in this lesson are:

☐ chicken thigh　☐ duck neck　☐ chicken breast　☐ quail
☐ goose web　☐ tender　☐ crisp　☐ fragrance
☐ recommend　☐ preparation　☐ procedure　☐ boil
☐ steam　☐ fry　☐ ingredients　☐ guard
☐ fried peanuts

I know __________ words.

Phrases and expressions I have learned in this lesson are:

☐ be famous for　☐ get rid of
☐ be worth of　☐ fry... until
☐ turn into　☐ hear of

I know __________ phrases and expressions.

I can:

☐ describe how to make the gourd chicken
☐ talk about the poultry

Note

Lesson 5 Livestock

扫码看课件

Goal

You will be able to:

1. Master the name of different livestock.
2. Know some cooking skills of livestock.

Warming up

❶ Read and match

Spare ribs	Beef	Streaky pork
Lamb	Pork	Beef shank

1. ________ 2. ________ 3. ________

4. ________ 5. ________ 6. ________

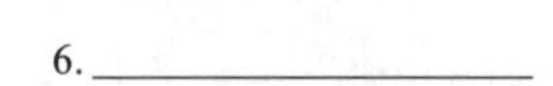

❷ Learn and say

A: What's your favorite meat?

B: My favorite meat is beef.

Let's learn

Section A

外教有声

❶ Listen to the dialogue and repeat

Lily: Bill, could you please teach me how to cook braised pork in brown sauce? I heard that it's quite delicious.

Note

Bill: My pleasure. This dish is really popular in China.

Lily: What do we need?

Bill: We need to prepare 500 g of cubed streaky pork, one cinnamon, three anise, five pieces of peeled ginger, two spoons of dark soy sauce, salt, and 15 g of sugar. Firstly, put some oil into the pot, and then put the cinnamon and anise into the oil, stir-fry them until the fragrance is smelled. Secondly, put the pork into the oil and add dark soy sauce, sugar and salt, mix the condiments well, and submerge the pork in boiled water after 5 minutes. Lastly, put in the ginger, and turn the low fire into high fire after 40 minutes.

Lily: Oh, it's really complex. Let's start with the streaky pork.

Bill: No problem.

❷ **Fill in the blanks with the verbs you have learned in this dialogue**

1.______ the pork

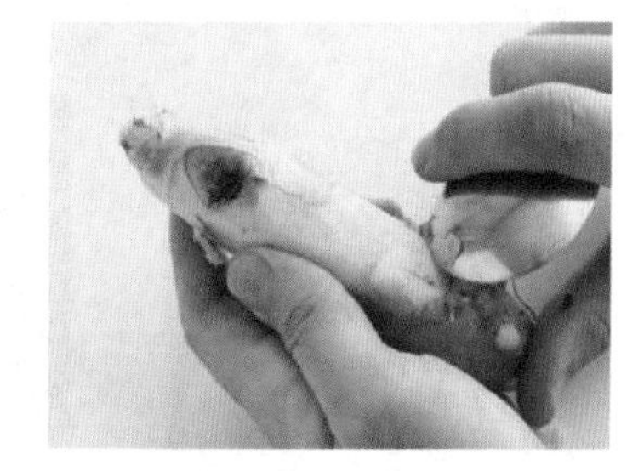

2.______ the ginger

3.______ the egg with tomato

4.______ the pork in the pot

❸ **Put the sentences below into the right order**

__________ Put the cinnamon and anise into the oil.

__________ Add dark soy sauce, sugar and salt.

__________ Put some oil into the pot.

__________ Turn the low fire into high fire.

__________ Put the pork into the oil.

__________ Submerge the pork in boiled water.

__________ Stir-fry them until the fragrance is smelled.

Section B

❶ **Read the short passage**

We often eat meat products in our daily life, including mutton, pork, beef and so on. The meat can be divided into white meat and red meat. They are divided by color before cooking. Red meat refers to pork, beef, mutton, venison and rabbit meat, etc. White meat generally refers to fish, shrimp, shellfish, chicken, duck and goose, etc. White meat is more greasy and soft. Red meat is

Note

harder and chewy, with less greasy. Children prefer to eat red meat, while the elderly choose white meat.

❷ **What meat can you see in the pictures? Please write them down**

1. ____________

2. ____________

3. ____________

4. ____________

5. ____________

6. ____________

❸ **Classify the meat in the above pictures**

Red meat: ____________________________

White meat: ____________________________

扫码看答案

Self-check

Words I have learned in this lesson are:

□ lamb	□ beef	□ pork
□ divide	□ chewy	□ greasy
□ soft	□ hard	□ cinnamon
□ anise		

I know ________ words.

Phrases and expressions I have learned in this lesson are:

□ beef shank	□ streaky pork	□ spare ribs
□ red meat	□ white meat	□ stir-fry
□ put... into...	□ start with	□ divide into
□ turn into	□ a spoon of	□ stir-fry... until...

I know ________ phrases and expressions.

I can:

□ describe how to make braised pork in brown sauce

□ classify the red meat and the white meat

扫码看课件

Lesson 6 Seafood

Goal

You will be able to:

1. Get familiar with different seafood.
2. Describe how to treat or cook seafood.

Warming up

1 Read and match

Oyster	Turtle	Eel	Fish	Crab
Shrimp	Octopus	Scallop	Conch	

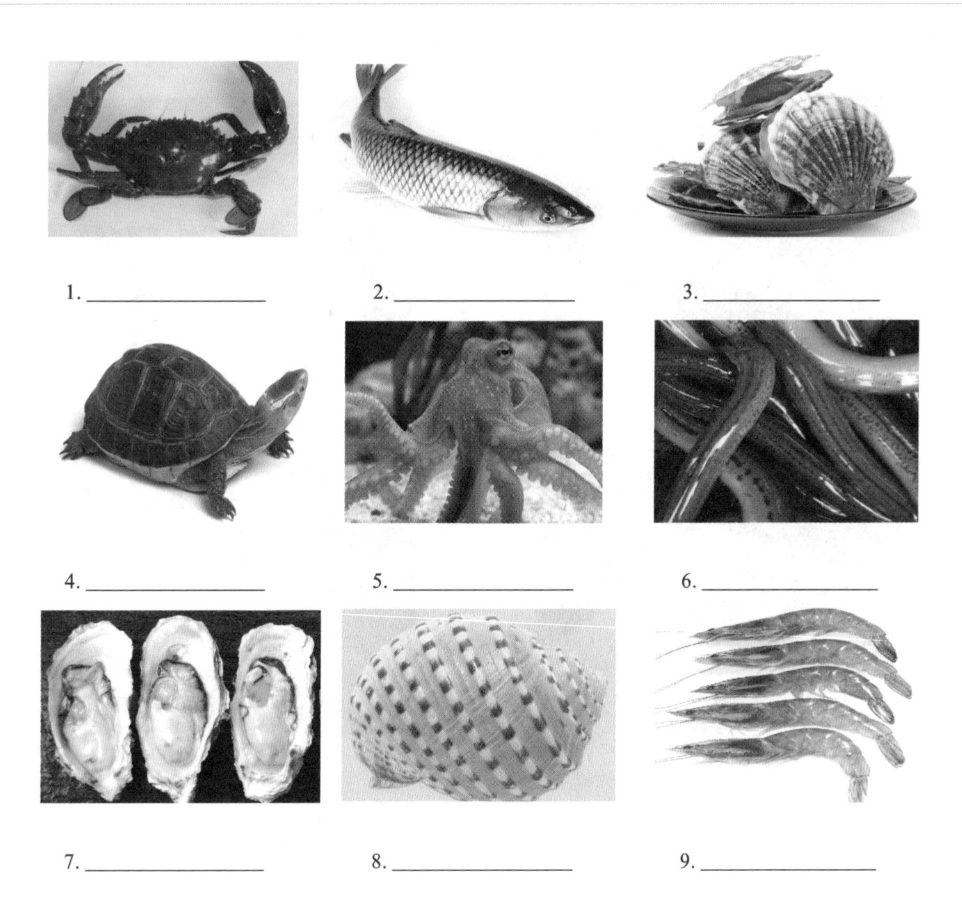

1. ________ 2. ________ 3. ________

4. ________ 5. ________ 6. ________

7. ________ 8. ________ 9. ________

2 Learn and say

A: What kind of seafood have you tasted?

B: I have tasted shrimps and crabs.

Let's learn

外教有声

Section A

❶ Listen to the dialogue and repeat 

Lily:Bill,guess what I bought for supper?

Bill:Wow,it's a fish,isn't it?

Lily:I don't know how to treat it before cooking. What shall we do?

Bill:Don't worry,Lily. I can do it.

Lily:What are we going to do at first?

Bill:First of all,clean the fish,cut it open,and then scale the fish carefully.

Lily:With what?

Bill:With a scaling knife,I have bought one in a hardware store.

Lily:What to do next?

Bill:Open it and take out the viscera and the gills. Last but not the least,remove the fish line.

Lily:That sounds a hard work. I give up.

Bill:Don't bother,I'll do it.

❷ Fill in the blanks with the verbs you have learned in this dialogue

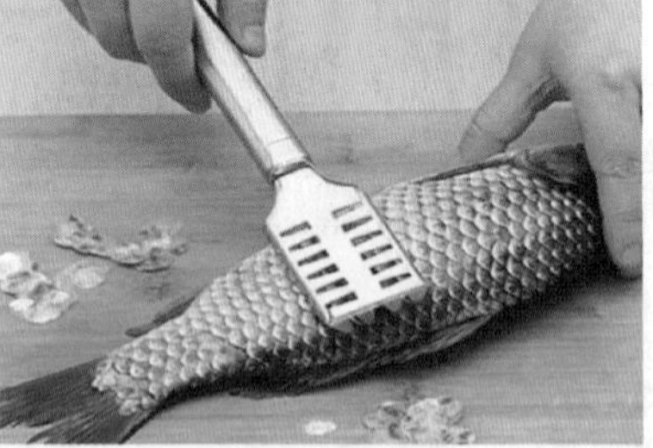

1.______the fish

2.______the fish

3.______the line of the shrimps

4.______the gills

❸ Decide true(T) or false(F)

(　　)1. Lily knows how to treat the fish before cooking.

(　　)2. First of all,They should clean the fish,and cut it open.

(　　)3. They are going to take out the viscera and the gills at last.

(　　)4. Bill bought the scaling knife in a supermarket.

(　　)5. Bill treats the fish at last.

Section B

❶ **Read the short passage**

The nutritive value of fish is extremely high. The fatty acids contained in fish promote the development of the brain. Eating fish is good for the brain of children. Fish improves their intellectual development. Fish also can prevent the elderly from suffering dementia. People eating fish regularly are likely to be more robust and have longer lifespans.

❷ **Match the pictures with the right words**

Bass	Grass carp	Catfish
Crucian	Hairtail	Yellow croaker

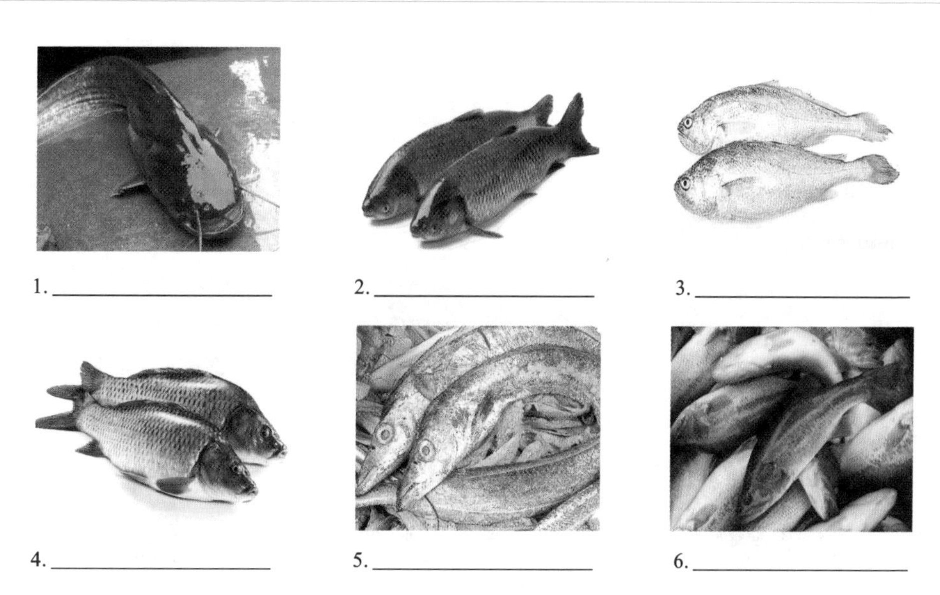

1. ____________ 2. ____________ 3. ____________

4. ____________ 5. ____________ 6. ____________

❸ **Talk with your partner about different kind of fish**

A: What kind of fish would you like to eat?

B: I would like to eat ____________.

Self-check

扫码看答案

Words I have learned in this lesson are:

□ turtle	□ eel	□ fish	□ crab
□ value	□ promote	□ improve	□ catfish
□ oyster	□ shrimp	□ bass	□ crucian
□ hairtail	□ scale	□ viscera	□ lifespan

I know ________ words.

Phrases and expressions I have learned in this lesson are:

□ yellow croaker	□ grass carp	□ fish line
□ be likely to	□ cut... open	□ give up
□ last but not the least		

I know ________ phrases and expressions.

Note

I can:

☐ describe how to treat seafood before cooking

☐ identify different kinds of seafood

Lesson 7 Condiments

扫码看课件

Goal

You will be able to:

1. Master the expressions of condiments.
2. Describe how to use the condiments.
3. Describe the different tastes of dishes.

Warming up

❶ Read and match

Scallion	Vinegar	Rice wine	Oil	Salt
Ginger	Soy sauce	Oyster oil	Garlic	

1. ____________ 2. ____________ 3. ____________

4. ____________ 5. ____________ 6. ____________

7. ____________ 8. ____________ 9. ____________

❷ Learn and say

A: What kinds of condiments do you often use during cooking?

B: We often use ____________.

Let's learn

Section A

❶ Listen to the dialogue and repeat

Lily:Bill,have you ever made hot pot?

Bill:Yes,I have learned how to make Chongqing hot pot last year,but it's really complex.

Lily:Why? I really want to have a try. Could you please teach me?

Bill:We need more than 15 condiments to make the hot pot seasoning.

Lily:Wow,what are they?

Bill: Let me think. Well, they are dry red chili, ginger, salt, garlic, Pixian soy bean paste, rapeseed oil,chili sauce,lard oil,cinnamon,prickly ash,and so on.

Lily:Wait a moment,I can't stuff so many words into my mind at once. Let me write them down.

Bill:Why don't we go shopping together? I will choose the condiments for you.

Lily:It couldn't be better.

❷ Fill in the blanks with the nouns you have learned in this dialogue

1.____________ 2.____________ 3.____________

4.____________ 5.____________ 6.____________

❸ Decide true(T) or false(F)

(　　)1. It is easy to make Chongqing hot pot.

(　　)2. You need more than 15 condiments to make the hot pot seasoning.

(　　)3. Garlic is not needed in the seasoning.

(　　)4. Dry red chili is needed in the seasoning.

(　　)5. Bill will help Lily to do shopping.

Section B

❶ Read the short passage

Benefits of vinegar

1. Relieve fatigue

Proper vinegar drinking can eliminate fatigue.

2. Balance acid-base in blood

Vinegar can mediate the balance of acid-base in blood.

3. Promote digestion

Vinegar can promote digestion and absorption. It can help people take in calcium effectively.

4. Prevent aging

The main component of vinegar is acetic acid, which has a strong bactericidal effect and can play an important role in protecting skin and hair.

5. Sterilization

Vinegar can enhance the bactericidal ability of intestines and stomach.

6. Protect liver and kidney

Vinegar can protect the liver and kidney.

❷ Match the pictures with proper words

❸ Read and practice the conversation

Lucy: I like to eat Mapo tofu. It's quite spicy but delicious. What about you?

Tom: I don't like spicy food, but I like something sweet or sour, such as sweet and sour fillet.

扫码看答案

Note

Self-check

Words I have learned in this lesson are:

☐ scallion ☐ oil ☐ salt ☐ ginger

☐ garlic ☐ cinnamon ☐ condiment ☐ fatigue
☐ promote ☐ digestion ☐ salty ☐ sour
☐ sweet ☐ bitter ☐ hot

I know __________ words.

Phrases and expressions I have learned in this lesson are:

☐ rice wine ☐ soy sauce ☐ oyster oil ☐ fry red chili
☐ soy bean paste ☐ rapeseed oil ☐ chili sauce ☐ lard oil
☐ prickly ash ☐ have a try ☐ stuff... into... ☐ easy to do
☐ at once ☐ more than ☐ take in

I know __________ phrases and expressions.

I can:

☐ name some common condiments
☐ describe the food I like

Note

Unit 3

Chinese Food

Lesson 8 Shandong Cuisine

扫码看课件

Goal

You will be able to:

1. Master the feature of Shandong cuisine.
2. Know some history of Shandong cuisine.
3. Learn and restate the process of making sweet and sour Yellow River carp.

Warming up

1 Read and match

Braised sea cucumber with scallion	Dezhou braised chicken
Sweet and sour Yellow River carp	Braised duck and pigeon
Braised intestines in brown sauce	Steamed tofu stuffed with vegetables

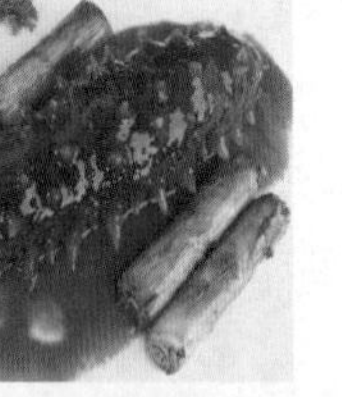

1. ____________

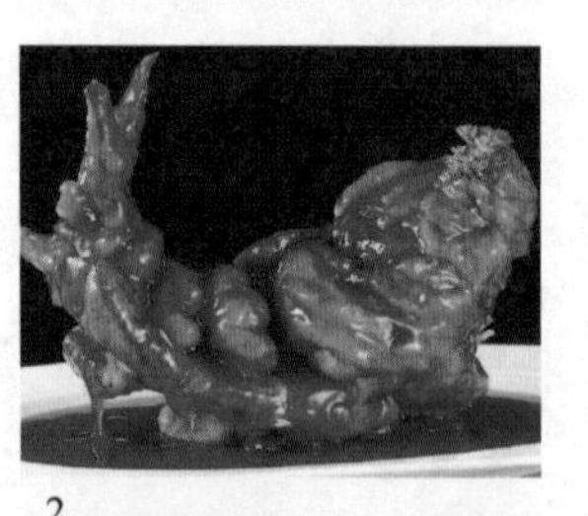

2. ____________

3. ____________

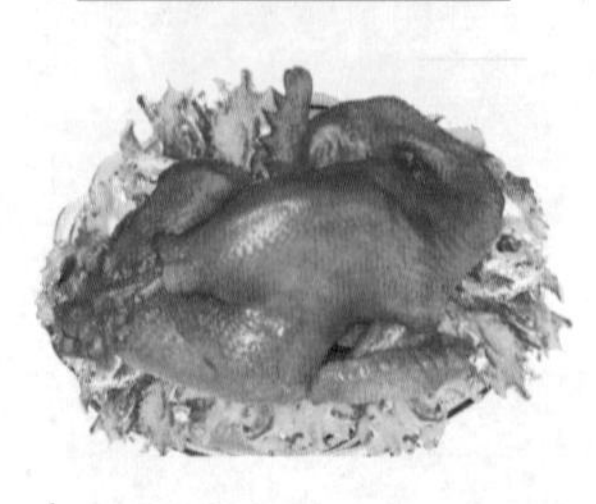

4. ____________

5. ____________

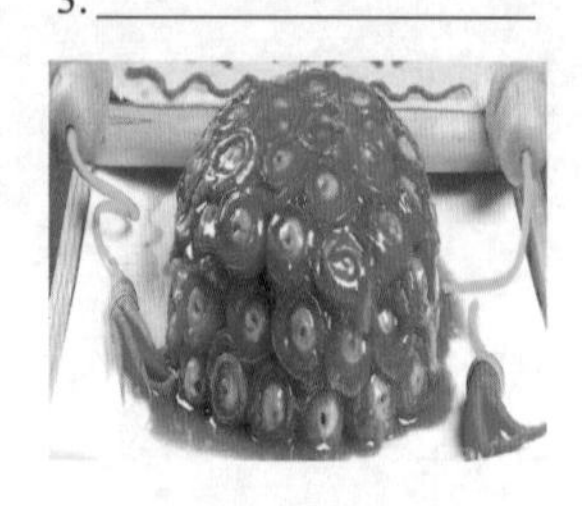

6. ____________

2 Learn and say

A: Do you like Shandong cuisine?

B: Yes, I do.

A: Can you name some Shandong dishes?

B: Dezhou braised chicken, sweet and sour Yellow River carp.

Section A

❶ Listen to the dialogue and repeat

Lily: Bill, I have been to Shandong recently. I find that I fall in love with Shandong cuisine now.

Bill: Really?

Lily: Shandong cuisine which is also named Lu cuisine, is one of the eight major cuisines in China. Its history can be dated back to the State of Qi in the Spring and Autumn period. However, because in the adjacent State of Lu, the great Confucius and Confucianism were born. Shandong cuisine is finally named after the State of Lu.

Bill: I see. What is the feature of Shandong cuisine?

Lily: Well, scallion and garlic are usually used as condiments. After thousands of years of development, Shandong cuisine is divided into many schools. For example, East Shandong school is good at cooking seafood, Confucius school pays great attention to luxury, Jinan school is particular about the degree of fire, while Zibo school is superb in making soup. Generations of cooks of Zibo have a say "gun is very important to soldiers and soup is to cooks".

Bill: Aha. Now I know something about Shandong cuisine.

❷ Answer the questions

1. Why is Shandong cuisine finally named after the State of Lu?
2. What are usually used as condiments of Shandong cuisine?

❸ Choose the best answer according to the dialogue

1. East Shandong school is good at ________

A. making soup　　B. cooking seafood　　C. the degree of fire

2. Confucius school pays great attention to ________

A. luxury　　B. the degree of fire　　C. soup

3. Jinan school is particular about ________

A. luxury　　B. the degree of fire　　C. seafood

4. Zibo school is superb in ________

A. making soup　　B. cooking seafood　　C. luxury

Section B

❶ Read the text with these questions in mind

1. What do you need to make sweet and sour Yellow River carp?
2. How to make sweet and sour Yellow River carp?

You need:

Main ingredient:

Yellow River carp

Condiments:

clear soup　soy sauce　cooking wine

vinegar　sugar　salt　wet starch

oil　onion　ginger　garlic

Steps:

Step 1. Cut the scales, viscera and cheeks of carp.

Step 2. Use a knife to cut the body of carp.

Step 3. Arrange clear soup, soy sauce, cooking wine, vinegar, sugar, salt and wet starch, and mix them thoroughly.

Step 4. Put the carp into wet starch and fry to seventy percent well, and then use soft fire to fry for 3 minutes until golden brown.

Step 5. Pour a little oil in wok, add chopped onion, ginger and garlic, and then put into well mixed sauce. Stir the sauce and pour it over the carp.

扫码看答案

Self-check

Words I have learned in this lesson are:

☐ braised sea cucumber with scallion

☐ sweet and sour Yellow River carp

☐ braised intestines in brown sauce

☐ Dezhou braised chicken

☐ steamed tofu stuffed with vegetables

☐ braised duck and pigeon

☐ adjacent　　☐ Confucius　　☐ Confucianism

I know ________ words.

Phrases and expressions I have learned in this lesson are:

☐ cut... into...　　☐ roll... into...

I know ________ phrases and expressions.

I can:

☐ name at least six representative dishes of Shandong cuisine

☐ describe the schools and feature of Shandong cuisine

☐ describe the procedure of cooking sweet and sour Yellow River carp

Lesson 9 Sichuan Cuisine

扫码看课件

Goal

You will be able to:

1. Master the expressions of Sichuan cuisine.
2. Describe how to make typical Sichuan dishes.

Warming up

❶ Read and match

Fried eggs with tomatoes	Yuxiang shredded pork
Mapo tofu	Twice-cooked pork slices
Kung pao chicken	Pork slices in spicy water

1.________ 2.________ 3.________

4.________ 5.________ 6.________

❷ **Learn and say**

A: Which dish do you like best?

B: I like ________ best.

A: Which cuisine does it belong to?

B: It belongs to Sichuan cuisine.

Let's learn

Section A

❶ **Listen to the dialogue and repeat**

外教有声

Lily: Bill, it's said that Mapo tofu is very delicious. Can you cook it?

Bill: Of course, I can.

Lily: Really? Can you teach me how to cook it?

Bill: Sure. First, cut the tofu into cubes, and drain the water from the tofu and set aside. Then heat a wok, and pour the oil and chili oil into it. Get this really hot and fry the pork mince. Remove with a slotted spoon but leave the oil behind. Add the bean paste and cook, stir for a few minutes until aromatic, and then add ginger and garlic. Add tofu cubes and pork mince to cook and stir for one minute or so, and then add in the wet starch flour and stir well for 20 seconds until it is a little sticky. Finally put into a plate and sprinkle over scallion and Sichuan peppercorns.

Lily: It sounds so tasty. Let's make it for dinner.

❷ **Fill in the blanks with the verbs you have learned in this dialogue**

1.____ with a slotted spoon

2.____ the oil into the wok

Note

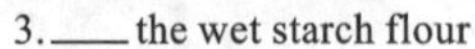
3.____ the wet starch flour

4.____ the tofu into cubes

Section B

❶ Read the short passage

Mapo tofu is a popular Chinese dish from Sichuan Province. The invention of Mapo tofu can be traced back to Qing Dynasty. Near Wanfu Bridge in Chengdu, there was a restaurant owned by a man named Chen whose wife had full of pockmarks on her face. She was called Chen Mapo. She developed a unique way of cooking tofu. The tofu that she made was quite delicious and not expensive. So it became popular and soon rose to fame. After the woman died, people named her specialty "Mapo tofu" in honor of this ordinary woman, because "Mapo tofu" literally means tofu made by the freckled woman. The dish has been passed on ever since and now becomes one of the representative dishes of Sichuan cuisine.

❷ Answer the questions according to the passage above

1. Which cuisine does Mapo tofu belong to?
2. Why is it called Mapo tofu?
3. When did Mapo tofu originate?

❸ Decide true (T) or false (F)

(　　)1. Mapo tofu belongs to Sichuan cuisine.
(　　)2. Mapo tofu originated in Ming Dynasty.
(　　)3. Chen Mapo became famous because the tofu she made tasted delicious with low price.

扫码看答案

Self-check

Words I have learned in this lesson are:

□ ginger　□ flavor　□ garlic　□ shredded
□ tofu cubes　□ starch flour　□ pork mince　□ slotted spoon
□ develop　□ unique　□ popular

I know ________ words.

Phrases and expressions I have learned in this lesson are:

□ drain ... from...　□ add in
□ cut... into...　□ rise to fame
□ representative dishes　□ Sichuan cuisine

I know ________ phrases and expressions.

I can:

□ retell the story of Mapo tofu
□ cook Mapo tofu

Lesson 10 Hunan Cuisine

扫码看课件

Goal

You will be able to:

1. Master the new words and expressions about Hunan cuisine.
2. Learn how to cook shredded pork with vegetables.
3. Learn some history of Hunan cuisine.

Warming up

❶ Read and match

Smoky flavors steamed together	Changsha stinky tofu
Spicy salted duck	Dong'an chicken
Mao's braised pork with soy sauce	Steamed fish head with chopped pepper

1. ____________ 2. ____________

3. ____________

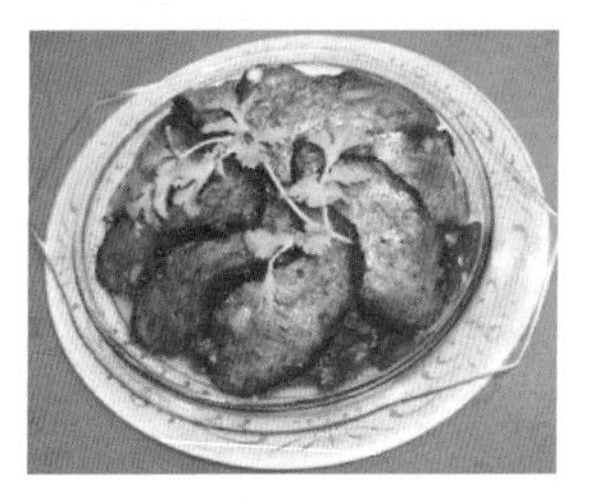

4. ____________ 5. ____________ 6. ____________

❷ Learn and practice

A: Do you like eating spicy food?

B: Yes, I like it.

A: Which spicy dish do you like best?

B: I like eating steamed fish head with chopped pepper best.

❸ Learn and act

hot	sweet	sour	fresh

Note

A: What taste do you like?

B: I like __________ food.

A: What spicy dishes do you have?

B: I have __________.

Let's learn

Section A

❶ Listen to the dialogue and repeat

外教有声

Lily: Bill, it's said that Hunan shredded pork with vegetables is very delicious. Can you cook it?

Bill: Of course, I can.

Lily: Really? Can you teach me how to cook it?

Bill: Sure. The steps are as follows:

Step 1. Wash and slice the peppers and the meat.

Step 2. Heat the pan without oil and then heat peppers over medium heat.

Step 3. Add oil and garlic into the pan and stir the garlic over medium heat until very fragrant.

Step 4. Add the streaky meat and keep stirring, and then add lean meat and stir them until they turn brown.

Step 5. Add stirred peppers and salt about 2 minutes later, and then add seasoned soy sauce for seafood and stir them for 2 minutes.

Lily: OK. I will have a try.

❷ Fill in the blanks with the words you have learned in this dialogue

1. Then, __________ lean meat and stir them until they turn brown.
2. __________ the garlic over medium heat until very fragrant.
3. __________ the pan without oil and then heat peppers over medium heat.
4. __________ and slice the peppers and the meat.
5. __________ stirred peppers and salt about 2 minutes later, and then add seasoned soy sauce for seafood and stir them for 2 minutes.

❸ Put the sentences above into the right order

1. __________ 2. __________ 3. __________ 4. __________ 5. __________

Section B

❶ Read the short passage

Hunan cuisine, sometimes called Xiang cuisine, is one of the eight major cuisines in China. The cooking skills employed in Hunan cuisine reached a high standard as early as the Western Han Dynasty, over 2100 years ago. Hunan cuisine consists of more than 4000 dishes, among which over 300 are very famous. It is characterized by its hot and sour flavor, fresh aroma, and deep color. It consists of three styles: Xiang River style, Dongting Lake style and western Hunan style.

Hunan cuisine is hot because the air is very humid which makes it difficult for the body to eliminate moisture. The local people eat hot peppers to help them remove dampness and coldness.

❷ Decide true (T) or false (F)

(　　) 1. Hunan cuisine is called Dongting cuisine.

(　　) 2. The cooking skills of Hunan cuisine have a history of over 2100 years.

( ) 3. Hunan cuisine is hot because the climate is very hot.

( ) 4. Hot peppers can't help remove dampness and coldness.

3 Answer the questions according to the passage above

1. When did the cooking skills of Hunan cuisine originate?
2. How many dishes does Hunan cuisine consist of?
3. What styles does Hunan cuisine have?

扫码看答案

Self-check

Words I have learned in this lesson are:

□ smoky	□ flavor	□ braised	□ chili
□ spicy	□ capital	□ typical	□ center
□ medium	□ pepper	□ fragrant	□ stir

I know ________ words.

Phrases and expressions I have learned in this lesson are:

□ spicy food	□ on business	□ the center of
□ heard of	□ stinky tofu	□ steamed fish head

□with chopped pepper

I know ________ phrases and expressions.

I can:

□ know the history of Hunan cuisine

□ know the famous traditional dishes of Hunan cuisine

Lesson 11 Zhejiang Cuisine

扫码看课件

Goal

You will be able to:

1. Master the expressions of cooking English.
2. Master some knowledge about Zhejiang cuisine.
3. Understand the origin of Dongpo pork.

Warming up

1 Read and match

West Lake fish in vinegar gravy   Fried shrimps with Longjing tea

West Lake water shield soup

Stir-fried beancurd rolls stuffed with minced tenderloin

Dongpo pork   Songsao shredded fish soup

Note

1. ____________

2. ____________

3. ____________

4. ____________

5. ____________

6. ____________

2 Learn and say

A: Have you ever had the famous Zhejiang dishes?

B: Yes, I have had Dongpo pork before.

Let's learn

Section A

外教有声

1 Listen to the dialogue and repeat 

Ben: Have you ever been to Zhejiang, Steven?

Steven: Yes. I've been there a few times.

Ben: Do you like the food there?

Steven: Very nice. I enjoy Zhejiang dishes very much.

Ben: I'm going to Hangzhou next week. Could you recommend some restaurants where I can eat typical Zhejiang dishes?

Steven: You can go to Louwailou restaurant, Tianwaitian restaurant, Waipojia restaurant, etc.

Ben: I've heard of Louwailou restaurant, which has a history of over 100 years. It's very famous for Zhejiang dishes.

Steven: Right. You can have authentic Zhejiang food there.

Ben: OK. I will have a try.

2 Complete the following dialogue and read it aloud

A: (1) ____________(你去过浙江吗), Steven?

B: Yes. I've been there (2) ________ (几次).

A: Do you like the food there?

B: (3) ________(很棒). I enjoy Zhejiang dishes very much.

A: (4) ____________(你能推荐一家饭店) where I can eat typical Zhejiang dishes?

B: You can go to Louwailou restaurant.

A: OK. I will have a try next week.

❸ Ask and answer the following questions about the dialogue with a partner

1. Where has Steven been?
2. Does Steven like the food of Zhejiang?
3. Why does Ben ask Steven to recommend the restaurants in Hangzhou to him?
4. Which restaurant is famous for Zhejiang dishes?
5. Will Ben go to that restaurant?

Section B

❶ Read the short passage

Dongpo pork

Dongpo pork is a famous dish in Zhejiang. It is red and bright in color, oily but not greasy, and crispy but not smashed with mellow juice. It's not only delicious, but also has an interesting origin.

There is an story about Dongpo pork. It is named after Su Dongpo, a well-known poet and artist in Song Dynasty. He is supposed to have invented this dish. In 1088 AD, the West Lake had not been renovated for a long time, and it was declining day by day. In this case, Su Dongpo went to Hangzhou and became the local official of there. He led the local people to dredge the West Lake, and built a dam across the lake. With the help of the dam, the local people had a good harvest each year.

The people of Hangzhou were very grateful, so they went to thank him with pork and rice wine. Su Dongpo accepted the gifts. And then he cooked the meat in his own way and shared it with everyone. The folks thought that it was very delicious and named the dish "Dongpo pork".

❷ Decide true (T) or false (F)

( ) 1. Dongpo pork is a famous dish in Hangzhou.
( ) 2. Dongpo pork is not only oily but also greasy.
( ) 3. Dongpo pork is named after Su Dongpo.
( ) 4. The local people cooked Dongpo pork to thank Su Dongpo.
( ) 5. Everyone thought that Dongpo pork was very delicious.

❸ Decide which dishes belong to Zhejiang cuisine

( ) 1. West Lake fish in vinegar gravy
( ) 2. Beggar's chicken
( ) 3. Fried shrimps with Longjing tea
( ) 4. Kung pao chicken
( ) 5. Songsao shredded fish soup
( ) 6. West Lake water shield soup

Self-check

扫码看答案

Words I have learned in this lesson are:

□ vinegar	□ shrimp	□ shredded	□ stuffed
□ tenderloin	□ typical	□ oily	□ greasy
□ mellow	□ poet	□ official	□ local
□ dam	□ renovate	□ authentic	

I know ________ words.

Phrases and expressions I have learned in this lesson are:

☐ name... after...　☐ be supposed to　☐ day by day

☐ be famous for　☐ in this case　☐ share... with...

I know ________ phrases and expressions.

I can:

☐ know the history of Dongpo pork

☐ know the famous traditional dishes of Zhejiang

Lesson 12 Jiangsu Cuisine

扫码看课件

Goal

You will be able to:

1. Master the expressions of cooking English.
2. Master some knowledge about Jiangsu cuisine.
3. Describe how to make stewed pork balls.

Warming up

1 Read and match

Stewed pork balls	Three-nested duck
Braised eels	Sweet and sour mandarin fish
Boiled salted duck	Braised shredded chicken with ham and dried tofu

❷ **Learn and say**

A: These dishes are famous in Jiangsu Province. Do you know them?

B: I know some of them.

Let's learn

Section A

❶ **Listen to the dialogue and repeat**

Ben: Steven, what do you think of Chinese food?

Steven: It's great. There are many kinds of delicious food in China.

Ben: What kind of food do you like?

Steven: I like sweet taste food.

Ben: Then you must like Jiangsu cuisine. Have you ever had stewed pork balls, a traditional dish of Jiangsu?

Steven: No. How is the stewed pork balls made?

Ben: It requires following materials: pork, water chestnuts, dried mushrooms, and some seasonings. First, chop the pork, water chestnuts and dried mushrooms. Put them into a container, add scallion and ginger water, salt, monosodium glutamate and starch, and then mix well. Next, make the pork balls. Boil for a short while, and then stew about 2 hours in the chicken soup with some sugar and soy sauce. It tastes soft, fat but not greasy. It is very delicious, and has rich nutrition.

Steven: Wow, I can't wait to try it.

❷ **Put the sentences below into the right order**

__________ Boil for a short while.

__________ Chop the pork.

__________ Make the pork balls.

__________ Mix the pork stuffing and the seasonings well.

__________ Stew the pork balls about 2 hours.

❸ **Ask and answer the following questions about the dialogue with a partner**

1. What kind of food does Steven like?
2. What is the taste of Jiangsu cuisine?
3. What raw materials do you need to make stewed pork balls?
4. Please briefly describe the steps of making stewed pork balls.
5. Do you know any other famous Jiangsu dishes?

Section B

❶ **Read the short passage**

Stewed pork balls

Stewed pork balls is a traditional dish in Huaiyang cuisine in Jiangsu Province. It is said that the stewed pork balls originated in Sui Dynasty. When Emperor Yang of the Sui Dynasty traveled along the Grand Canal, he was fascinated by the beautiful scenery of Yangzhou. When he returned to the palace, he asked the royal chefs to make dishes according to the beautiful scenery of Yangzhou. With the help of Yangzhou chefs, the royal chefs made four dishes with all their thoughts. The four dishes are sweet and sour mandarin fish, shrimp cake, stir chicken strips with tender flower stalk, and pork meatballs. The emperor was very happy after tasting the four dishes.

Note

So he gave a banquet to all the ministers. Everyone was attracted by the dishes. Huaiyang cuisine became popular for a while.

The pork meatballs is the predecessor of the stewed pork balls. Control of heat is very important in making this dish. Stewing about 40 minutes with weak fire. In this way, the pork balls are fat but not greasy, and very soft and tender. Today, stewed pork balls is one of the famous dishes of Jiangsu cuisine. You can find it in every Jiangsu style restaurant.

❷ Decide true (T) or false (F)

(　　) 1. Stewed pork balls is made of pork.

(　　) 2. Stewed pork balls is not only oily but also greasy.

(　　) 3. Stewed pork balls is a traditional Jiangsu dish.

(　　) 4. The chef cooked stewed pork balls according to an interesting story.

(　　) 5. The ministers thought stewed pork balls was very delicious.

❸ Decide which dishes belong to Jiangsu cuisine

(　　) 1. Boiled salted duck

(　　) 2. Mapo tofu

(　　) 3. Sweet and sour mandarin fish

(　　) 4. Stewed pork balls

(　　) 5. Beggar's chicken

(　　) 6. Braised shredded chicken with ham and dried tofu

扫码看答案

Self-check

Words I have learned in this lesson are:

☐ eel　☐ boil　☐ traditional　☐ material

☐ seasoning　☐ scallion　☐ monosodium glutamate　☐ starch

☐ soy sauce　☐ nutrition　☐ stew　☐ dynasty

☐ emperor　☐ banquet　☐ predecessor

I know ________ words.

Phrases and expressions I have learned in this lesson are:

☐ originate in　☐ it is said　☐ be fascinated by

☐ according to　☐ with the help of　☐ be attracted by

I know ________ phrases and expressions.

I can:

☐ know the history of stewed pork balls

☐ know the famous traditional dishes of Jiangsu

Lesson 13 Fujian Cuisine

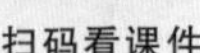

扫码看课件

Goal

You will be able to:

1. Master the feature of Fujian cuisine.

2. Know some history of Fujian cuisine.
3. Learn and restate the process of making litchi meat.

Warming up

1 Read and match

Buddha jumping over the wall	Seven-star fish ball
Red wine fish steak	White snow chicken
Fujian peanuts	Litchi meat

1. ________ 2. ________ 3. ________

4. ________ 5. ________ 6. ________

2 Learn and say

A:Do you like Fujian cuisine?

B:Yes,I do.

A:Can you name some Fujian dishes?

B:Buddha jumping over the wall,Fujian peanuts.

Let's learn

Section A

1 Listen to the dialogue and repeat 

外教有声

Lily:What shall we have for lunch,Bill?

Bill:I'm hungry. Let's go for a bite to eat.

Lily:What about Shaxian delicacies? It's a kind of Fujian cuisine.

Bill:Good idea! Can you tell me something about Fujian cuisine?

Lily:Fujian cuisine,also called Mincai for short,is one of the eight major cuisines in China,and originated from Minhou County in Fuzhou City. It had begun to take shape in the Tang Dynasty and Song Dynasty.

Bill:I see. What is the feature of Fujian cuisine?

Lily: Well, Fujian cuisine comprises three branches: Fuzhou cuisine, South Fujian cuisine and West Fujian cuisine. Fujian cuisine is mainly based on Fuzhou cuisine, which is fresh and lite and somewhat sweet and sour. Represented by Xiamen flavor, cuisine in the south of Fujian is fresh, sweet and spicy. Represented by Changting cuisine, the Hakka dishes in the west of Fujian are greasy and salty.

Bill: Aha. Let's go and have a big meal of Fujian cuisine.

❷ Answer the questions

1. Where did Fujian cuisine originate?

2. How many branches does Fujian cuisine comprise?

❸ Choose the best answer according to the dialogue

1. Fujian cuisine, __________ Mincai for short, is one of the eight major cuisines in China.

A. is called B. also called C. are called

2. Fujian cuisine is mainly based on __________.

A. Fuzhou B. South Fujian C. West Fujian

3. The feature of Xiamen flavor is __________.

A. greasy and salty B. fresh and lite and somewhat sweet and sour

C. fresh, sweet and spicy

4. The feature of the Hakka dishes is __________.

A. greasy and salty B. fresh and lite and somewhat sweet and sour

C. fresh, sweet and spicy

Section B

Read the text with these questions in mind

1. What do you need to make litchi meat?

2. How to make litchi meat?

You need:

Main ingredients:

300 grams of pork	100 grams of water chestnuts

Condiments:

chicken gravy	lees
vinegar	garlic
one egg white	25 grams of ginger
10 grams of white sugar	15 grams of corn starch

Steps:

Step 1. Cut the meat into cruciferous shape and then cut them into triangles.

Step 2. Cut the water chestnuts into slice.

Step 3. Add some chicken gravy, lees, egg white and corn starch, and stir them together.

Step 4. Make the slide of meat with cruciferous shape face outside, roll it into a round, and place it neatly.

Step 5. Make sauce with vinegar, sugar, garlic and other condiments.

Step 6. Fry the litchi meat that is shaped.

Step 7. Add some sauce and stir them evenly.

扫码看答案

Self-check

Words I have learned in this lesson are:

□ Buddha jumping over the wall	□ seven-star fish ball
□ red wine fish steak	□ white snow chicken
□ Fujian	□ peanuts
□ litchi meat	□ originated
□ delicacies	□ flavor greasy

I know ________ words.

Phrases and expressions I have learned in this lesson are:

□ cut... into...	□ roll... into...

I know ________ phrases and expressions.

I can:

□ name at least six representative dishes of Fujian cuisine

□ describe the feature of Fujian cuisine

□ restate the process of making litchi meat

Lesson 14 Cantonese Cuisine

扫码看课件

Goal

You will be able to:

1. Master the new words and expressions about Cantonese cuisine.
2. Learn how to cook Cantonese soup.
3. Learn some history of Cantonese cuisine.

Warming up

1 Read and match

Spring roll	Boiled prawns	Shrimp dumpling
Roasted crispy suckling pig	Steamed rice roll	Sliced boiled chicken

1. ________________

2. ________________

3. ________________

4.____________ 5.____________ 6.____________

❷ Learn and practice

A: Do you like eating Cantonese food?

B: Yes, I like it.

A: What Cantonese dishes do you like best?

B: I like eating roasted goose and sliced boiled chicken.

❸ Learn and act

Cantonese cuisine	Hunan cuisine	Shandong cuisine	Fujian cuisine

A: May I take your order now?

B: Yes. I'd like to try some Cantonese dishes. Could you recommend some?

A: OK. How about canton soup and roasted goose.

B: It sounds good. I'll take them.

Let's learn

Section A

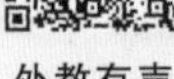
外教有声

❶ Listen to the passage and repeat

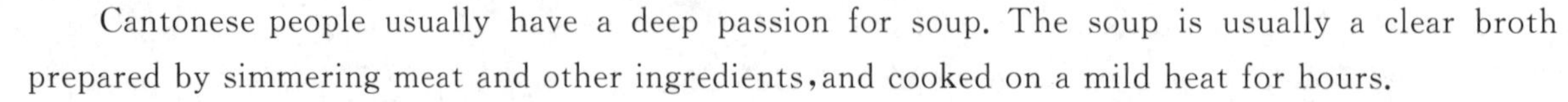

Cantonese people usually have a deep passion for soup. The soup is usually a clear broth prepared by simmering meat and other ingredients, and cooked on a mild heat for hours.

Unlike other Chinese cuisine, the Cantonese usually serve soup before a meal. The soup-drinking tradition is said to be related to the regional climate in Guangdong, which is moist heat.

The Cantonese soup refers to the soup that is cooked for a long time and tasty. It is usually cooked with the clay pot. First, put the raw materials like ribs or chicken in the pot, when the water is boiling, put the fire down and cook it with slow fire for 2 to 4 hours. There are hundreds of recipes of the soup. The turtle soup, pigeon soup, shark's fin soup, abalone soup and some others are popular. Since 1990s, all the restaurants of Guangzhou serve tasty soup. Some of the restaurants offer telephone-reserve service.

❷ Decide true (T) or false (F)

(   ) 1. Cantonese soup is cooked on a mild heat for hours.

(   ) 2. Cantonese usually serve soup after a meal.

(   ) 3. The climate in Guangdong is moist heat.

(   ) 4. The clay pot isn't important for the tasty soup.

(   ) 5. The turtle soup, pigeon soup, shark's fin soup, abalone soup and some others are popular.

❸ Answer the questions

1. What is soup-drinking tradition in Guangdong related to?

2. When is the soup served in a meal?

3. What pot do Cantonese people use to cook the soup?

4. What popular Cantonese soups are mentioned in the passage?

Section B

❶ Read the short passage

Yum Cha(饮茶) is an ancient tradition in Guangdong. When they meet in the morning, they usually greet each other with "Have you had tea?"

The tea-drinking tradition can be traced back to the Qing Dynasty. In those days, the Cantonese used to go to a nearby teahouse where they needed to pay only two cents for a pot of tea and some simple snacks.

With the passage of time, morning tea becomes a custom that provides diverse population of Guangdong with a common cultural identity: Cantonese. And in the morning tea, in addition to having tea, they also have all kinds of dishes and snacks.

Drinking morning tea has become a life habit for people living in Guangdong. Drinking tea is a mode of social exchange, which is an important factor for the long history of prosperous teahouses in Guangdong through the centuries.

Every morning, the Cantonese come to drink morning tea for different reasons. The real tea drinkers, for instance, prefer to kill time with hot tea and snacks. Businessmen came here to exchange information as well as to enjoy life. But thousands of ordinary people would rush to the teahouse in the early morning for a moment of relaxation before starting their daily work.

❷ Decide true (T) or false (F)

( ) 1. Drinking morning tea can be traced back to the Ming Dynasty.

( ) 2. The Cantonese used to go to a teahouse for a pot of tea with some snacks.

( ) 3. Drinking morning tea has become a lifestyle for Cantonese.

( ) 4. People drink morning tea because they are thirsty.

❸ Answer the questions according to the passage above

1. How do people in Guangdong greet each other in the early morning?

2. When can the tea-drinking tradition be traced back to?

3. What do Cantonese have in the morning tea now?

4. Why do many ordinary people come to teahouse now?

Self-check

扫码看答案

Words I have learned in this lesson are:

□ climate	□ regional	□ moist
□ broth	□ mild	□ recipes
□ abalone	□ prosperous	□ relaxation

I know ________ words.

Phrases and expressions I have learned in this lesson are:

□ spring roll	□ boiled prawns	□ sliced boiled chicken
□ roasted crispy suckling pig	□ shrimp dumpling	
□ steamed rice roll	□ refer to	

I know ________ phrases and expressions.

I can:

☐ know the history of Cantonese cuisine

☐ know the famous traditional dishes of Cantonese cuisine

Lesson 15 Anhui Cuisine

扫码看课件

Goal

You will be able to:

1. Master the feature of Anhui cuisine.
2. Know the history of Anhui cuisine.
3. Know how to make smelly mandarin fish.

Warming up

1 Read and match

Li Hongzhang hotchpotch	Fu Liji braised chicken
Stewed soft shell turtle with ham	
Bamboo shoots cooked with sausage and dried mushroom	
Tunxi rice wine-steeped crab	Smelly mandarin fish

1. ____________

2. ____________

3. ____________

4. ____________

5. ____________

6. ____________

2 Learn and say

A: Which dish do you like best?

B: I like ________ best.

A: Which cuisine does it belong to?

B: It belongs to ________ cuisine.

Let's learn

Section A

1 Listen to the dialogue and repeat

外教有声

Lily: Bill, I have been to Anhui recently. I find that I fall in love with Anhui cuisine now.

Bill: Really? Would you like to tell me something about it?

Lily: I'd like to. Anhui cuisine, one of the eight major cuisines in China, includes the south of Anhui cuisine, Huaibei cuisine and Yanjiang cuisine.

Bill: I see. How about its history?

Lily: Oh, it originated in a small town named She at the foot of Yellow Mountain in Han Dynasty, and it developed in Song Dynasty and Yuan Dynasty, and then flourished in Ming Dynasty and Qing Dynasty.

Bill: And what is the feature of Anhui cuisine?

Lily: It is elegant and simple, original taste, crisp and tender. Anhui cuisine chefs pay more attention to the taste and color of dishes and the timing of cooking, and they are good at braising and stewing.

Bill: Aha. Now I like Anhui cuisine some.

2 Answer the questions

1. Is Anhui cuisine one of the eight major cuisines in China?
2. Where did Anhui cuisine originate?
3. When did Anhui cuisine develop?
4. What are the Anhui cuisine chefs good at?

3 Choose the best answer according to the dialogue

1. I ________ Anhui cuisine now.

A. am interested in B. fall in love with C. am good at

2. The Anhui cuisine includes ________.

A. the south of Anhui cuisine

B. Huaibei cuisine and Yanjiang cuisine

C. both A and B

3. Anhui cuisine chefs ________ the taste, color of dishes and the temperature ________ them.

A. pay more attention to, to cook

B. pay less attention to, to cook

C. pay more attention to, cook to

4. Anhui cuisine chefs ________ braising and stewing.

A. do well at B. are good at C. are good in

Section B

1 Read the text with these questions in mind

1. What do you need to make smelly mandarin fish?
2. Is it easy or difficult to make smelly mandarin fish?

Note

3. How to make smelly mandarin fish?

You need:

Main ingredient:

750 grams of mandarin fish

Auxiliary ingredients:

50 grams of streaky pork	50 grams of winter bamboo shoots

condiments:

50 grams of soy sauce	10 grams of white chestnut powder
25 grams of ginger	10 grams of white sugar
25 grams of leek	15 grams of cooking wine

Steps:

Step 1. Rub salt onto the mandarin fish, and salted for six or seven days.

Step 2. Cut off the scale and open the stomach, remove everything inside. Then wash the fish in clean water and dry it.

Step 3. Cut the pork and bamboo shoots into slices and cut the leek into small sections.

Step 4. Put the fish into the hot oil, and then take it out until both sides are golden brown.

Step 5. Fry the sliced streaky pork and sliced bamboo shoots, and then put in the fried fish and the condiments.

Step 6. Add corn starch solution and dish up.

❷ Put the following pictures into the right order

1. ____________ 2. ____________ 3. ____________

4. ____________ 5. ____________ 6. ____________

❸ Imitate and practice

Cut the leek into small sections.

Cut the pork and bamboo shoots into slices.

Cut off the scale and open the stomach, remove everything inside.

Cut down the tomato to the skin but leaving the skin intact.

Words I have learned in this lesson are:

☐ attention ☐ braise ☐ crisp ☐ dynasty

☐ elegant ☐ hotchpotch ☐ introduce ☐ intact

☐ leek ☐ mandarin ☐ originate ☐ rub

☐ remove ☐ section ☐ slice ☐ smelly

☐ temperature ☐ tender ☐ turtle

I know ________ words.

Phrases and expressions I have learned in this lesson are:

☐ at the foot of ☐ be good at...

☐ cut...into... ☐ fall in love with

☐ pay attention to ... ☐ put...into...

☐ dish up

I know ________ phrases and expressions.

I can:

☐ know something about Anhui cuisine

☐ describe how to make smelly mandarin fish

Unit 4

Foreign Food

Lesson 16 Japanese Cuisine

扫码看课件

Goal

You will be able to:

1. Remember the English words of Japanese cuisine.
2. Know the ingredients in ramen.
3. Know the method of cooking miso soup.

Warming up

1 Read and match

Tempura	Sashimi	Sushi
Miso soup	Ramen	Kobe steak

1. ____________________

2. ____________________

3. ____________________

4. ____________________

5. ____________________

6. ____________________

2 Learn and say

A: What are you making?

B: I'm making a __________ for dinner.

A: Do you need a hand?

Note

B: Sure. That would be nice. You can wash the vegetable that for the ________. Thanks.

Let's learn

Section A

❶ Listen to the dialogue and repeat 

Lily: Bill, do you want to go out for dinner tonight?

Bill: No, I am going to cook dinner at home. Would you like to come over to eat?

Lily: Yes, thank you for inviting me. What are you going to fix for dinner?

Bill: I am going to make ramen.

Lily: Oh, really? OK, that's fantastic. How are you going to do ramen?

Bill: Ramen are made with wheat flour, water, salt and kansui. But it's hard for me to make it. So I will choose the instant ramen noodles. And I will use fresh chicken wings for the broth. It takes much more time if I use fresh chicken wings, but that will make the broth different from others.

Lily: That sounds good. Do you need any help making dinner or do you want me to bring something?

Bill: No, just come to my house at 6:00 pm.

Lily: OK, I'll see you tonight.

❷ Answer the questions

1. What kind of food are Bill cooking tonight?
2. What ingredients are in ramen?
3. What are ramen broth made of?
4. When do they meet?

❸ Choose the best answer according to the dialogue

1. Do you want to ________ dinner tonight?

A. go out to　　B. go out for　　C. go out

2. Would you like to ________ to eat?

A. come over　　B. come on　　C. come by

3. It takes ________ time if I use fresh chicken wings.

A. many　　B. more much　　C. much more

4. Do you need any help ________ dinner?

A. make　　B. making　　C. made

Section B

Read the text with these questions in mind

1. Is it easy or difficult to make miso soup?
2. How to make miso soup?

You need:

Main ingredient:

4 cups of water

Auxiliary ingredients:

1 (8 ounces) package silken tofu, diced

2 green onions, diagonally sliced into 1/2 inch pieces

Seasonings:

2 teaspoons of dashi granules　　3 tablespoons of miso paste

Note

Steps:

Step 1. In a medium saucepan over medium-high heat, combine dashi granules and water; bring to a boil.

Step 2. Reduce heat to medium, and whisk in the miso paste. Stir in tofu.

Step 3. Separate the layers of the green onions, and add them to the soup.

Step 4. Simmer gently for 2 to 3 minutes before serving.

扫码看答案

Self-check

Words I have learned in this lesson are:

□ Japanese	□ tempura	□ sashimi	□ sushi
□ miso soup	□ ramen	□ Kobe steak	□ fantastic
□ wheat flour	□ kansui	□ instant	□ broth
□ ounce	□ package	□ silken	□ diced
□ onion	□ sliced	□ diagonally	□ inch
□ piece	□ teaspoon	□ tablespoon	□ paste
□ medium	□ saucepan	□ combine	□ separate
□ simmer	□ gently		

I know ________ words.

Phrases and expressions I have learned in this lesson are:

□ go out for	□ come over
□ fix for dinner	□ be made with...
□ make... different from others	□ sliced... into...
□ bring to a boil	□ reduce heat to
□ whisk in	□ stir in
□ add... to	□ simmer for

I know ________ phrases and expressions.

I can:

□ describe how to make miso soup

□ describe what ingredients are in ramen

□ describe the famous food in Japan

Lesson 17 Korean Cuisine

扫码看课件

Goal

You will be able to:

1. Remember the English words of Korean cuisine.
2. Know the ingredients in Korean barbecued chicken.
3. Know the method of cooking grilled ribs.

Warming up

❶ Read and match

Grilled ribs	Grilled pork belly	Korean scallion pancake
Kimchi	Korean barbecued chicken	Ginseng chicken soup

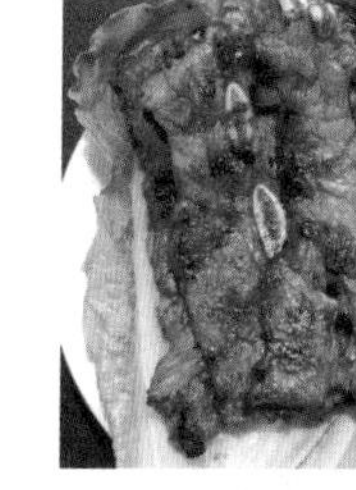

1.____________ 2.____________ 3.____________

4.____________ 5.____________ 6.____________

❷ Learn and say

A: I'm cooking __________. Do you want some?

B: Yes, please.

A: Do you like __________? I'm going to cook for dinner.

B: Great! I like it.

Let's learn

Section A

❶ Listen to the dialogue and repeat

外教有声

Chef: Today we will use the beef short ribs to make this new dish—grilled ribs.

Commis cook: Woo, it sounds very delicious.

Chef: Yes, it is. Please cut the short ribs into pieces first.

Commis cook: No problem. And then?

Chef: Place the ribs in a large stockpot and cover with cold water. Soak ribs in refrigerator for one hour to pull out blood. Drain.

Commis cook: I'm finished. I think I will now use something for marinating ribs.

Chef: You are right. We can combine garlic, onion, and Asian pear in a blender and puree. Pour into a large bowl, and stir in the soy sauce, brown sugar, honey, sesame oil and black pepper. Then marinate ribs in the soy mixture.

Commis cook: For how long?

Chef: Better to marinate overnight, but you can also marinate for only six hours.

Commis cook: OK, I have got it.

Six hours later...

Commis cook: It has been six hours, chef. What should I do next?

Chef: Now we can grill ribs until the meat is tender and the outside is crusty, about 5 to 10 minutes per side.

Commis cook: The ribs are done. Should I garnish it?

Chef: I have prepared some lettuce. You can use them for garnish.

❷ Answer the questions

1. What kind of food are chef cooking today?
2. What ingredients are in grilled ribs?
3. How to marinate ribs?
4. How long do they marinate ribs?

❸ Choose the best answer according to the dialogue

1. Place the ribs ________ a large stockpot.

A. in B. down C. on

2. Soak ribs in refrigerator for one hour to ________ blood.

A. pull B. pull out C. pull out of

3. I think I will now use something ________ marinating ribs.

A. to B. as C. for

4. Now we can grill ribs ________ the meat is tender.

A. with B. for C. until

Section B

Read the text with these questions in mind

1. Is it easy or difficult to make Korean barbecued chicken?
2. How to make Korean barbecued chicken?

You need:

Main ingredient:

200 grams of boneless and skinless chicken thighs

Auxiliary ingredient:

1 teaspoon of roasted sesame seeds

Marinade:

1/2 cup of sesame oil	3 tablespoons of soy sauce
1/3 cup of shallots, finely chopped	4 teaspoons of minced garlic
4 teaspoons of mirin	1/2 teaspoon of salt
2 teaspoons of Korean red chile flakes	1/2 teaspoon of pepper
3 tablespoons of light corn syrup	
4 teaspoons of finely grated peeled fresh ginger	

Steps:

Step 1. Rinse the chicken thighs, remove the skin and bones. Pat dry with paper towels and set aside.

Step 2. In a large bowl, mix all the ingredients of the marinade together.

Step 3. Add the chicken thighs into the marinade, make sure to stir coat the chicken thighs evenly. Cover and refrigerate for two hours.

Step 4. Fire up the grill, brush a bit of oil on the surface. Grill the chicken thighs until they turn golden brown on both sides. Remove the chicken thighs from grill and cut crosswise into 1/2-inch thick strips. Serve immediately with the alongside.

扫码看答案

Self-check

Words I have learned in this lesson are:

□ Korean □ kimchi □ grilled □ rib
□ thick □ barbecue □ alongside □ strip
□ stockpot □ cover □ soak □ refrigerated
□ drain □ marinate □ blender □ crosswise
□ puree □ sesame □ mixture □ overnight
□ tender □ crusty □ garnish □ boneless
□ skinless □ thigh □ shallot □ chopped
□ minced □ mirin □ chile □ syrup
□ grated □ rinse □ remove □ pat
□ grill □ brush □ surface □ immediately

I know ________ words.

Phrases and expressions I have learned in this lesson are:

□ grilled pork belly □ Ginseng chicken soup
□ Korean scallion pancake □ Korean barbecued chicken
□ use... to do... □ cut... into...
□ cover with □ pull out
□ use... for doing... □ got it
□ per side □ pat dry
□ set aside □ make sure to do
□ fire up □ turn golden brown

I know ________ phrases and expressions.

I can:

□ describe how to make grilled ribs
□ describe what ingredients are in Korean barbecued chicken
□ describe how to make Korean barbecued chicken

Lesson 18 Southeast Asian Cuisine

扫码看课件

Goal

You will be able to:

1. Remember the English words of Southeast Asian cuisine.
2. Apply the expressions into conversation.
3. Know the method of cooking chicken satay.

Warming up

1 Read and match

Bak Kut Teh	Singapore noodles	Tom yum
Curry chicken	Hainanese chicken rice	Satay

1. ________ 2. ________ 3. ________

4. ________ 5. ________ 6. ________

2 Learn and say

A: Have you made the ________?

B: I'm ready to cook the next dish.

A: Are you going to make a ________?

B: Yes.

Let's learn

Section A

1 Listen to the dialogue and repeat 

外教有声

Commis cook: What are we going to do?

Chef: Now we are making curry chicken. Can you prepare the chicken thighs first?

Commis cook: Yes, how many chicken thighs do we need?

Chef: 500 grams, please. Cube them first.

Commis cook: OK.

Chef: Sprinkle the cubes with salt, and set them aside for 15 minutes. Meanwhile let's prepare potatoes, onions and garlic.

Commis cook: Yes, I will crush 3 cloves of garlic. How about onions and potatoes?

Chef: Peel the onions and potatoes. Then cut them into pieces.

Commis cook: No problem.

Chef: I think the salted chicken thighs are done. Can you take them here?

Commis cook: OK, done.

Chef: Now I will show you how to cook the curry chicken. First, heat oil in a skillet over medium-high heat. Cook garlic, onions and potatoes for 3 minutes until onions are translucent.

Commis cook: Yes, I will cook them until onions are translucent.

Chef: Next, add chicken thighs and cook until they change from pink to white. Add curry powder and cook for 2 minutes.

Commis cook: Sure. I will do that.

Chef: Now add coconut milk and chicken stock. Stir, low heat to medium heat and cook, simmering rapidly, for 10 minutes until sauce reduces and thickens.

Commis cook: Is that finished?

Chef: No, you can taste to see if it needs more salt. Garnish with coriander.

Commis cook: OK, chef.

❷ Answer the questions

1. What kind of food are chef cooking?
2. What ingredients are in curry chicken?
3. How to make curry chicken?
4. How does the curry sauce look like?

❸ Choose the best answer according to the dialogue

1. Set them ________ for 15 minutes.

A. off　　B. aside　　C. down

2. Cook garlic, onions and potatoes for 3 minutes ________ onions are translucent.

A. for　　B. to　　C. until

3. Add chicken thighs and cook until they change ________ pink to white.

A. to　　B. from　　C. while

4. You can taste to see if it needs ________ salt.

A. more　　B. many　　C. much

Section B

Read the text with these questions in mind

1. How long should we marinade the satay?
2. How to make chicken satay?

You need:

Main ingredient:

300 grams of boneless and skinless chicken thighs

Auxiliary ingredients:

5 bamboo skewers, soaked in cold water for 2 hours

1 cucumber, cut into small pieces

1 small onion, quartered

oil, for basting

Marinade:

2 stalks of lemon grass, white parts only

2 cloves of garlic, peeled

2 teaspoons of turmeric powder

1 teaspoon of coriander powder

1/2 tablespoon of salt or more to taste

3 tablespoons of oil

6 shallots, peeled

1 teaspoon of chili powder

2 tablespoons of sugar

Steps:

Step 1. Cut the chicken thighs into small cubes. Set aside.

Step 2. Blend all the marinade ingredients in a food processor. Add a little water if needed.

Step 3. Combine the chicken and the marinade together, stir to mix well. Marinate the chicken for 6 hours in the fridge, or best overnight. When ready, thread 3-4 pieces of the chicken with the bamboo skewers.

Step 4. Grill the chicken satay skewers for 2-3 minutes on each side until the chicken is fully cooked and the surface is nicely charred on both sides. Baste and brush with some oil while grilling. Serve hot with the fresh cucumber pieces and onions.

扫码看答案

Self-check

Words I have learned in this lesson are:

□ southeast	□ Asian	□ Bak Kut Teh	□ Singapore
□ Tom yum	□ curry	□ Hainanese	□ satay
□ commis	□ thigh	□ cube	□ meanwhile
□ crush	□ clove	□ peel	□ heat
□ skillet	□ medium	□ translucent	□ powder
□ coconut	□ simmer	□ rapidly	□ reduce
□ thicken	□ taste	□ garnish	□ coriander
□ marinade	□ boneless	□ skinless	□ bamboo
□ skewer	□ soak	□ basting	□ stalk
□ lemon grass	□ shallot	□ turmeric	□ chili
□ blend	□ processor	□ combine	□ fridge
□ overnight	□ thread	□ grill	□ surface
□ baste	□ brush		

I know ________ words.

Phrases and expressions I have learned in this lesson are:

□ set... aside for	□ cut... into...
□ show... how to...	□ heat... in...
□ over medium-high heat	□ change from... to...
□ simmering rapidly	□ each side
□ both sides	

I know ________ phrases and expressions.

I can:

□ describe how to make curry chicken

□ describe what ingredients are in chicken satay

□ describe how to make chicken satay

Lesson 19 Turkish Cuisine

扫码看课件

Goal

Note

You will be able to:

1. Remember the English words of Turkish cuisine.
2. Know the ingredients and the steps of making Turkish pide.
3. Know something about Turkish doner kebab.

Warming up

❶ Read and match

Grilled fish	Doner kebab	Cappadocia crock beef
Turkish dumpling	Turkish pide	Dolma

1.__________ 2.__________ 3.__________

4.__________ 5.__________ 6.__________

❷ Learn and say

A:What's your favorite Turkish dish?

B:I like ________ best. How about you?

A:I like ________. Are you getting used to the food here?

B:I'm not really getting used to it yet.

Let's learn

Section A

❶ Listen to the dialogue and repeat

外教有声

Lily:Bill,chef Arslan taught me how to make Turkish pide yesterday.

Bill:Oh,great! What ingredients does it need?

Lily: All-purpose flour, dry active yeast, olive oil, milk, lemon juice, minced meat, peeled tomatoes,paprika,onion,parsley,salt,minced garlic and so on.

Bill:It sounds so complex.

Lily:I think so,and the steps are also complicated. Here are the steps.

Bill: Oh, it's really not an easy job.

Steps for making Turkish pide

Step 1. Mix all-purpose flour, dry active yeast, olive oil, warm water and salt to a dough, ferment for 60 minutes until the dough double in size.

Step 2. Punch down the dough and divide it into two portions.

Step 3. Roll the dough out the traditional paddle shape. Fold the sides over and pinched the ends, and then brush the dough with milk.

Step 4. Mix all the ingredients for the topping and spread over the prepared dough evenly.

Step 5. Bake in a preheated 190 ℃/375 F oven for 25 minutes. Sprinkle the pide with the lemon juice.

❷ Listen again and try to memorize the ingredients

1. ________ 2. ________ 3. ________

4. ________ 5. ________ 6. ________

❸ Read the list of steps and put them into the right order

________ Punch down the dough and divide it into two portions.

________ Mix all-purpose flour, dry active yeast, olive oil, warm water and salt to a dough.

________ Sprinkle the pide with the lemon juice.

________ Ferment the dough for 60 minutes until the dough double in size.

________ Bake in a preheated 190 ℃/375 F oven for 25 minutes.

________ Roll the dough out the traditional paddle shape. Fold the sides over and pinched the ends.

________ Brush the dough with milk.

________ Mix all the ingredients for the topping.

________ Spread over the prepared dough evenly.

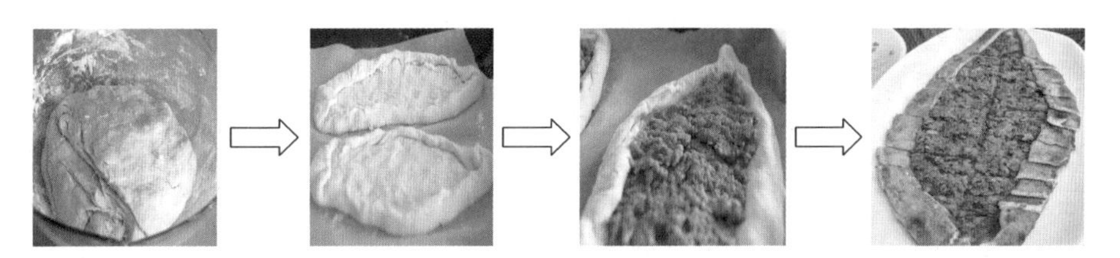

Section B

❶ Read the text with these questions in mind

1. What does doner kebab literally mean?

2. What is the authentic Turkish way to enjoy a kebab?

Turkish doner kebab

You may have seen signs that say "Turkish doner kebab" while in China or just about anywhere else in the world. But today, why not travel to Turkey with us, and experience the most authentic Turkish snacks on the streets of Istanbul?

"Doner kebab" literally means "rotating kebab" in Turkish. Chunks of lamb, beef or chicken on a stick rotate in the vertical position over an open fire. Though people around the world eat doner kebab with different side dishes such as French fries or salads, the authentic Turkish way to enjoy a kebab is to wrap the sliced meat in pita bread with tomatoes, onions, spices and mayonnaise. It is recognized as Turkish people's favorite snack.

❷ Decide true(T) or false(F)

(　　)1. You may have seen signs that say "Turkish doner kebab" all over the world.

(　　)2. Lamb, beef and rabbit are the main ingredients of doner kebab.

(　　)3. The authentic Turkish way to enjoy a kebab is to wrap the sliced meat in pita bread with tomatoes, onions, spices and blueberry jam.

(　　)4. Doner kebab is recognized as Turkish people's favorite snack.

❸ Match the pictures with right answer

1.________ 2.________ 3.________

4.________ 5.________ 6.________

A. mayonnaise	B. spices	C. tomatoes
D. pita bread	E. onions	F. doner kebab

扫码看答案

Self-check

Words I have learned in this lesson are:

□ authentic	□ crock	□ dolma	□ dumpling
□ dough	□ flour	□ grill	□ kebab
□ mayonnaise	□ mince	□ paprika	□ parsley
□ pinch	□ rotate	□ snack	□ sprinkle
□ topping	□ vertical	□ wrap	□ yeast

I know __________ words.

Phrases and expressions I have learned in this lesson are:

□ all-purpose flour　□ Cappadocia crock beef
□ doner kebab　□ dry active yeast
□ minced garlic　□ mix... to...
□ Turkish pide

I know __________ phrases and expressions.

I can:

□ know something about Turkish cuisine
□ describe how to make Turkish pide

Lesson 20 Western Cuisine

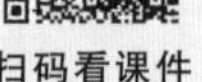
扫码看课件

Goal

You will be able to:

1. Remember the English words of Western cuisine.
2. Know some cooking methods of Western cuisine.
3. Know the degrees of steak doneness.

Warming up

① Read and match

Pasta	Foie gras	Beef tartare
Sandwich	Roast turkey	Caviar

1. __________________

2. __________________

3. __________________

4. ____________ 5. ____________ 6. ____________

❷ Learn and say

A: I've finished making ________. Is there anything else you need?

B: Could you grate some cheese?

A: Sure. Which dish are you going to use for this cheese?

B: ________.

Let's learn

Section A

❶ Listen to the dialogue and repeat

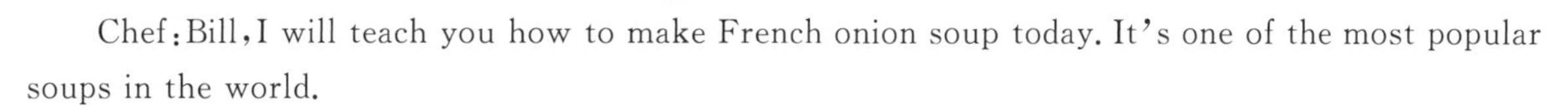

Chef: Bill, I will teach you how to make French onion soup today. It's one of the most popular soups in the world.

Commis cook: Yes, I can't wait to learn how to make it.

Chef: The main ingredients for this soup are onions, butter, beef broth, cheese and French bread. Alex has sliced ten onions yesterday. You can just cook them in butter.

Commis cook: Over low heat?

Chef: Yes, please, stir constantly.

Commis cook: For how long?

Chef: About 20 minutes, and then you can add the beef broth and water. Also put some bay leaves, pepper and thyme in it to add flavor.

Commis cook: OK, done. Shall I heat the soup to a boil now?

Chef: Fine, but remember to lower the heat when the soup is boiling. Then cover the pan and keep at a simmer for about 20 minutes. In the meantime, you can toast the French bread. Then pour the soup into bowls and top each with a slice of toasted French bread.

Commis cook: Is that all?

Chef: No, the final step is to sprinkle grated cheese on top of the soup and put the bowls in the salamander just before serving the soup to keep warm.

Commis cook: OK, I have got it. Thank you!

Chef: Not at all.

❷ Answer the questions

1. What is French onion soup?
2. How to make French onion soup?
3. How can we add flavor to the onion soup?
4. Why put the bowls in the salamander just before serving the soup?

Note

③ Choose the best answer according to the dialogue

1. It's ________ popular soups in the world.

A. one of the most　　B. very among　　C. the most

2. Remember to ________ the heat when the soup is boiling.

A. lower　　B. reduce　　C. both A and B

3. Then ________ the soup ________ bowls and ________ each ________ a slice of toasted French bread.

A. pour, in, top, to　　B. pour, to, top, into　　C. pour, into, top, with

4. The final step is to sprinkle grated cheese ________ the soup.

A. in top of　　B. on top of　　C. for top of

Section B

Read the text with these questions in mind

1. What kind of degrees of steak doneness do you like?

2. How to make pan fried beef steak with black pepper sauce?

You need:

Main ingredient:

200 grams of beef rib eye steak

Auxiliary ingredients:

2 stalks of fresh green asparagus	2 grams of flour
5 milliliter of soya oil	1 gram of garlic, peeled
14 grams of butter	20 grams of mango

Seasonings:

3 grams of chicken stock powder	3 grams of ground black pepper
14 grams of light soy sauce	14 grams of tomato ketchup
2 grams of fruity sauce	3 grams of sugar

1 tablespoon of chopped fresh mango (optional)

Steps:

Step 1. Marinate the beef in a mixture of soya oil, ground black pepper, fruity sauce and flour. Set aside.

Step 2. Sauté the asparagus and add the garlic, chicken stock powder and flour. Pan fry the marinated beef to the degrees of steak doneness that customers like. The degrees of steak doneness contains blue, rare, medium rare, medium, medium well, well done.

Step 3. Make the sauce, melt butter, and add tomato ketchup, light soy sauce, chicken stock powder and sugar. Mix well.

Step 4. Place a small amount of black pepper sauce on a plate before arranging the steak over. Garnish the plate with the asparagus and chopped fresh mango (optional).

扫码看答案

Note

Self-check

Words I have learned in this lesson are:

☐ western　☐ pasta　☐ foie gras　☐ beef tartare

☐ sandwich　☐ roast turkey　☐ caviar　☐ grate

- ☐ French
- ☐ onion
- ☐ commis
- ☐ broth
- ☐ cheese
- ☐ slice
- ☐ stir
- ☐ constantly
- ☐ bay
- ☐ thyme
- ☐ flavor
- ☐ pour
- ☐ simmer
- ☐ sprinkle
- ☐ salamander
- ☐ stalk
- ☐ peel
- ☐ marinate
- ☐ chop
- ☐ sauté
- ☐ pan fry
- ☐ garnish

I know ________ words.

Phrases and expressions I have learned in this lesson are:

- ☐ use for
- ☐ how to make
- ☐ can't wait to
- ☐ main ingredients for... are...
- ☐ heat... to a boil
- ☐ lower the heat
- ☐ in the meantime
- ☐ pour... into...
- ☐ top... with...
- ☐ put... in...
- ☐ set aside
- ☐ mix well
- ☐ garnish the plate with...

I know ________ phrases and expressions.

I can:

- ☐ describe how to make French onion soup
- ☐ describe how to make pan fried beef steak with black pepper sauce
- ☐ describe the degrees of steak doneness

Note

Unit 5

Pastries

Lesson 21 Chinese Pastries

扫码看课件

Goal

You will be able to:

1. Master the expressions of Chinese pastries.
2. Describe how to make Chinese pastries.
3. Introduce some kinds of Chinese pastries to others.

Warming up

1 Read and match

Dumpling	Moon cake	Mung bean cake
Pumpkin pie	Pancake	Sachima
Spring roll	Rice pudding	Steamed buns

1. ________

2. ________

3. ________

4. ________

5. ________

6. ________

7. ______

8. ______

9. ______

❷ Learn and say

A: What kind of Chinese pastries would you like?

B: I would like to have some steamed buns.

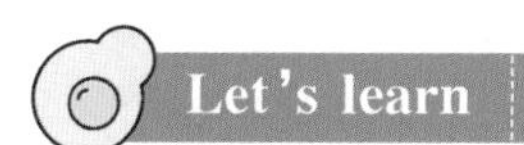

Section A

Dialogue A

❶ Listen to the dialogue and repeat 

Lily: Bill, is this your first time to China?

Bill: Yes, I like China so much, especially the Chinese food.

Lily: Which do you like best?

Bill: Dumplings of course. It is said that it is a traditional food in China, and they call it Jiaozi.

Lily: Wow. You know a lot about Jiaozi. Do you know why does Chinese people have it on Spring festival?

Bill: I'm afraid I have no idea about that. Can you tell me?

Lily: Because Chinese people think the pronunciation of Jiaozi is the same as "jiaozi", which means to ring out the old year and ring in the new year. And Jiaozi also symbolizes family reunion in Chinese people's mind.

Bill: It is marvelous.

Lily: Come and have Jiaozi with me. My treat.

Bill: I'd love to. Thank you very much.

❷ Match A on the left with B on the right

Lantern festival	Jiaozi
Dragon boat festival	sweet dumplings
Spring festival	moon cake
Mid-autumn festival	Zongzi

❸ Make dialogues with your partners

moon cake	rice dumplings wrapped in reed leaves
Mid-autumn festival	Dragon boat festival
family reunion	to commemorate Qu Yuan

Dialogue B

❶ Listen to the dialogue and repeat 

Zhang Yuan: Hey, Kate. I would like to give you a surprise.

Note

Kate: Oh, what surprise?

Zhang Yuan: I took some Chinese pastries for you. You told me that you have been expecting for a long time.

Kate: Thank you so much for your kindness. Can I have a try?

Zhang Yuan: Certainly. Help yourself.

(After a while)

Zhang Yuan: How do you feel about them?

Kate: Delicious! What is the name of this one? I think this kind is the best.

Zhang Yuan: Its name is Mung bean cake. It is a traditional Chinese pastries.

Kate: How to make it?

Zhang Yuan: First, you should steep the mung beans in water for about 12 hours, and then remove the peel. Second, steam the beans for 30 minutes, and after that, mill the beans into mud. Third, put some butter and oil in a wok, when the butter is melted, add the beans and fry them for 5 minutes, and then add sugar and continue to fry till the mixture becomes a flour dough. Finally, take a little dough, roll it into a ball, and press it into the shapes which you like with mould. You can also put them in your fridge, and that would be more tasty.

Kate: My dear! It's wonderful, but I'd rather buy some than do it!

Zhang Yuan: I think so.

❷ Fill in the blanks with the verbs you have learned in this dialogue

1.______ the mung beans in water

2.______ it for 30 minutes

3.______ the mixture

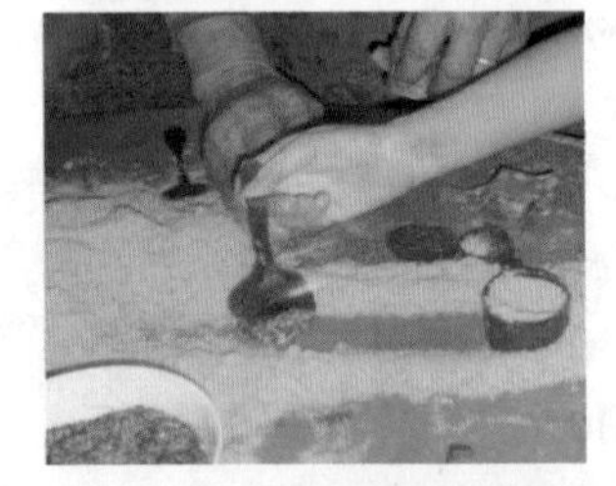

4.______ the beans into mud

5.______ it into a ball

6.______ it into the shapes

❸ Put the sentences below into the right order

__________ Take a little dough, roll it into a ball.

__________ Fry the mixture to become a flour dough.

__________ Steam the beans for 30 minutes, and then mill the beans into mud.

__________ Steep the mung beans in water for about 12 hours, and then remove the peel.

__________ Press it into the shapes with mould.

Section B

Passage A

❶ Read the short passage

Chinese people like to eat sweet dumplings on the 15th of lunar new year. Do you know how to make them? Let me tell you the steps. First, you can mix the glutinous rice flour with warm water, and then knead the mixture into dough till it has a smooth surface, leave it for several minutes. Second, prepare some sugar, sesame powder, lard, etc., mix them in a container. Then we get the filling. Third, Take a little dough, rub it into a piece, put the filling in it, wrap and knead the dough into a ball. At last, put the dough balls into the boiling water, and boil them till the balls float. It tastes better when they are hot.

❷ Translate the following into Chinese

1. Mix the flour with water.
2. Knead the mixture into dough.
3. Take a little dough, rub it into a piece.
4. Put the mixed filling in it, wrap and knead the dough into a ball.
5. Put the dough balls into the boiling water, and boil them till the balls float.

❸ Fill in the blanks with the verbs you have learned in this passage

1.______ the flour with water

2.______ it into dough

3.______ some mixed filling

4.______ it into a piece

5.______ the dough into a ball

6.______ the dough balls into the boiling water

Passage B

❶ Read the short passage

In China, people like all kinds of Chinese pastries. Chinese people like to call them "dim sum". Chinese dim sum has a long history. It is said that there was a general in the Eastern Jin Dynasty. His soldiers were very brave and fought bravely. He was moved by the soldiers, so he ordered the chef to make delicious cakes and pastries which were popular with the citizens. Then he sent the

pastries to the battlefield to comfort the soldiers.

The pastries represent the general's appreciation, which is the same pronunciation as "diandianxinyi" in Chinese. From then on, the word "dim sum" is spread, and now Chinese people still use it.

❷ Decide true (T) or false (F)

(　　)1. Chinese people like dim sum.

(　　)2. The name of dim sum was invented by a chef.

(　　)3. The general was in Tang Dynasty.

(　　)4. The general sent the dim sum to his soldiers.

❸ Discuss with your partner: what kind of dim sum do you like best? Why?

扫码看答案

Self-check

Words I have learned in this lesson are:

□ dumpling	□ mung bean	□ pudding	□ bun
□ traditional	□ symbolize	□ reunion	□ marvelous
□ pastry	□ steep	□ remove	□ steam
□ mill	□ melt	□ dough	□ rub
□ mould	□ knead	□ sesame	□ wrap
□ float	□ chef		

I know ＿＿＿＿＿＿ words.

Phrases and expressions I have learned in this lesson are:

□ Spring festival	□ Chinese pastries
□ rub... into...	□ hot to serve
□ knead... into...	□ fry... till
□ be popular with	□ lunar new year
□ dim sum	□ help yourself

I know ＿＿＿＿＿＿ phrases and expressions

I can:

□ describe how to make sweet dumplings and mung bean cake

□ have some knowledge about Chinese culture

Lesson 22 Western Pastries

扫码看课件

Goal

You will be able to:

1. Master the expressions of western pastries.
2. Describe how to make western pastries.
3. Introduce some kinds of western pastries to others.

Warming up

❶ Read and match

Cake	Puff	Sandwich	Tart	Doughnut
Bread	Macaroon	Apple pie	Tiramisu	

1. ______________

2. ______________

3. ______________

4. ______________

5. ______________

6. ______________

7. ______________

8. ______________

9. ______________

❷ Learn and say

A: What would you like for dessert?

B: I'd like to have a piece of chocolate cake.

Let's learn

Section A

Dialogue A

❶ Listen to the dialogue and repeat 

Waiter: Is everything all right with your dinner?

Guest: Yes, everything is good. Thank you.

Waiter: What would you like for the dessert?

Guest: Can you give me some suggestions?

外教有声

Note

Waiter: Certainly. We have cake, bread, pudding, pies and egg tarts, etc. The egg tarts is our chef's recommendation.

Guest: I would like to have some egg tarts.

Waiter: What do you think of our tarts?

Guest: It's tasty. Is it easy to make?

Waiter: Yes. First, put some milk and light cream in a clean container, and then add some sugar and stir to blend. Second, prepare another clean container, add egg yolk, and sift into the low gluten flour, stir well. Third, add the mixture into the egg yolk paste, stir it. Fourth, pour the mixed liquid into the puff pastry. Finally, preheat the oven to 200 ℃ for 10 minutes, and then put the egg tarts into the oven to bake them for about 20 minutes till the tarts turn brown. You can have a try.

Guest: Oh, that sounds interesting. I will try.

❷ Fill in the blanks with the verbs you have learned in this dialogue

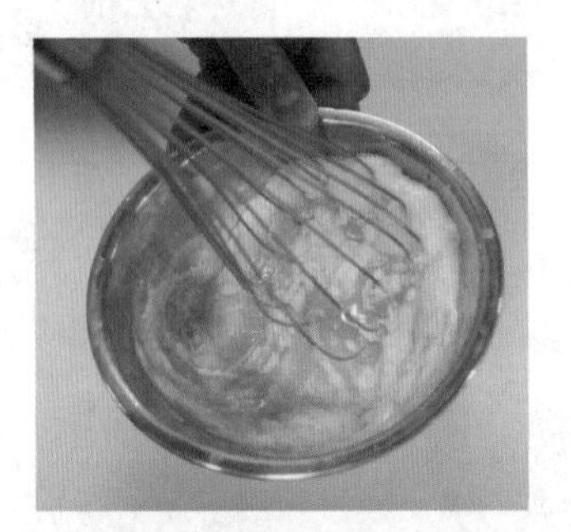

1.______ to blend

2.______ into the low gluten flour

3.______ them for about 20 minutes

4.______ the oven to 200 ℃

❸ Put the sentences below into the right order

__________ Bake them for about 20 minutes till the tarts turn brown.

__________ Put some milk and light cream in a clean container.

__________ Add egg yolk, and sift into the low gluten flour, stir well.

__________ Pour the mixed liquid into the puff pastry.

__________ Add the mixture into the egg yolk paste, stir it.

Dialogue B

❶ Listen to the dialogue and repeat

Kevin: Hi, Jenny. Would you like to try some desserts?

Jenny: Yes. What kind of dessert?

Kevin: Tiramisu. It's a kind of Italian cake.

Jenny: What does it mean?

Kevin: It's about a romantic story. It is said that once in Italy, there was a soldier who would go into battle. He was so poor, but his wife loved him very much. His wife prepared food for him when he was about to leave, but she found there was nothing to eat. So she made a cake with the bread and biscuits all they have. She called the cake "tiramisu". It means that "to take me away" and "remember me". Whenever the soldier ate the cake, he would remember his family and his wife.

Jenny: It's quite special. I can't wait to taste it.

Kevin: What do you think of the cake?

Jenny: It tastes fragrant, smooth and sweet. I like it.

❷ Read again and fill in the blanks

1. Tiramisu is a kind of __________ cake.
2. It tastes __________, __________ and sweet.
3. The __________ would remember his family and his wife when he ate the cake.
4. The story is so __________.

❸ Put the sentences into English

1. 你要吃点甜点吗？
2. 这是一种意式甜点。
3. 当士兵吃蛋糕时，他就会想起自己的家庭和妻子。
4. 我都等不及品尝它了。
5. 它尝起来香滑甜腻。

Section B

Passage A

❶ Read the short passage

Every year we celebrate our birthdays with cake, but do you know why people have it on birthdays? It is said that the first people to start celebrating the birthdays were the Ancient Romans. But the Germans were the first people to celebrate birthdays with cake for their children. In 18th century, the Germans celebrated kids' birthdays with cake and candles. The birthday cake was an important part of the celebration. And every child was given a cake with one candle for every year they had lived, plus one extra to symbolize the hope of living another year. When blowing out the candles, every child had to make a wish, just as what we still do today.

❷ Decide true (T) or false (F)

(　　)1. We usually have birthday cake on New Year's Day.

(　　)2. The first people to start celebrating birthdays were the Germans.

(　　)3. The Germans were the first people to celebrate birthdays with cake.

(　　)4. When blowing out the candles, every child had to sing a song in 18th century.

❸ Discuss in groups and talk about Chinese customs on birthday

Passage B

❶ Read the short passage

Do you know how to make delicious cookies? Let's have a try! First, put the sugar and butter into a container, stir the mixture quickly about 20 minutes till it turns to white. Second, add an egg into the mixture, and then mix well. Third, add the low-gluten flour, stir well. Fourth, put the mixture into a pastry bag, squeeze it into the shape which you like. At last, put the cookies in oven, 150 ℃ for about 15 minutes. Bake the cookies till the color of the edge changes. You can add different ingredients according to your own taste, such as nuts, green tea powder, chocolates and so on. Baking is difficult but interesting, you should be patient.

❷ Fill in the blanks with the verbs you have learned in this passage

1. ______ the sugar and butter ______ a container.
2. ______ the mixture quickly about 20 minutes ______ it turns to white.
3. ______ an egg ______ the mixture.
4. ______ it ______ the shape which you like.

❸ Put the sentences below into the right order

______ Stir the mixture quickly about 20 minutes till it turns to white.
______ Put the mixture into a pastry bag, squeeze it into the shape which you like.
______ Add the low-gluten flour, stir well.
______ Add an egg into the mixture, and then mix well.
______ Bake the cookies till the color of the edge changes.
______ Put the sugar and butter into a container.

扫码看答案

Self-check

Words I have learned in this lesson are:

□ tart	□ doughnut	□ macaroon	□ pie
□ tiramisu	□ dessert	□ suggestions	□ container
□ recommendation	□ add	□ stir	□ egg yolk
□ low gluten flour	□ sift	□ puff pastry	□ preheat
□ oven	□ biscuits	□ fragrant	□ smooth
□ celebrate	□ kid	□ extra	□ squeeze
□ ingredients	□ patient	□ romantic	□ pastry bag

I know ______ words.

Phrases and expressions I have learned in this lesson are:

□ stir to blend	□ have a try
□ add... into...	□ bake... till...
□ it is said that	□ put... into...
□ prepare... for...	□ can't wait to
□ celebrate... with...	□ blow out
□ green tea powder	□ turn to

I know ______ phrases and expressions.

I can:

□ describe how to make egg tart and cookies
□ get some knowledge about the history of cake and tiramisu

Unit 6

Cuisine Matching and Decorating

Lesson 23 Cuisine Matching

扫码看课件

Goal

You will be able to:

1. Identify types of cuisine.
2. Make cuisine matching.

Warming up

1 Read and match

Fish   Tofu   Mutton   Spinach   Tofu & fish soup   Chinese chestnut
Stewed chicken & Chinese chestnut   Clam   Egg   Steamed clams and egg
Chicken   Scallion   Sautéed sliced lamb with scallion   Liver   Spinach liver soup

1. ________________

2. ________________

3. ________________

4. ________________

5. ________________

6. ________________

7. ________________

8. ________________

9. ________________

❷ Learn and say

A: What dish do you like best?

B: I like tofu & fish soup.

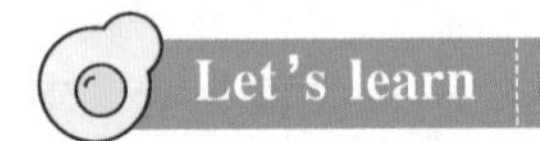

Let's learn

Section A

❶ Listen to the dialogue and repeat 

外教有声

Lily: Bill, what are you doing?

Bill: I'm trying to make a dish named sautéed sliced lamb with scallion.

Lily: It's a popular Chinese dish. Would you please teach me to do it?

Bill: Of course. First, wash the lamb, scallion, pepper and your fried pot. Then, try to slice the lamb and scallion into the same shape. Next, pour some oil into the dry pot. Fry the lamb over the high heat for 8 minutes.

Lily: Hmm, it smells tasted. I can't wait to taste it.

Bill: It isn't the time. We haven't finished yet. We need to put scallion into the pot and fry them with the lamb, add 2 spoons of water, and cook for 2 minutes. Add a little bit salt. It's done.

Lily: Wow, what a delicious food it is!

Bill: Now, let's enjoy it.

❷ Fill in the blanks with the verbs you have learned in this dialogue

1. ________ oil

2. ________ pot

Note

3.________ salt

4.________ pepper

5.________ mutton

6.________ dish

❸ Put the sentences below into the right order

__________ Then, try to slice the lamb and scallion into the same shape.

__________ First, wash the lamb, scallion, pepper and your fried pot.

__________ Fry the lamb over the high heat for 8 minutes.

__________ Put scallion into the pot and fry them with the lamb.

__________ Next, pour some oil into the dry pot.

__________ Finally, add a little bit salt.

Section B

❶ Read the short passage

Cuisine matching is very important in cooking. It includes color matching, shape matching, flavor matching, nutrition matching and so on. While cooking, a cook should pay attention to the above tips. He or she should not only learn cuisine matching theory knowledge but also master cooking skills. Only in this way can he or she gain in the long run.

❷ Please write down the cuisine matching

1.____________________

2.____________________

3.____________________

4.____________________

扫码看答案

Self-check

Words I have learned in this lesson are:

□ clam □ fish □ liver □ mutton
□ oil □ pepper □ pot □ pour
□ salt □ scallion □ slice □ spinach
□ tasted □ tofu

I know ________ words.

Phrases and expressions I have learned in this lesson are:

□ pay attention to
□ put... into...
□ sautéed sliced lamb with scallion
□ stewed chicken & Chinese chestnut
□ steamed clams and egg
□ tofu & fish soup

I know ________ phrases and expressions.

I can:

□ identify types of cuisine
□ make cuisine matching

Lesson 24 Cuisine Decoration

扫码看课件

Goal

You will be able to:

1. Identify types of cuisine decoration.
2. Learn requirements of cuisine decoration.
3. Make cuisine decoration.

Warming up

❶ **Read and match**

Vegetable decoration	Jam picture decoration	Paste decoration
Plate decoration	Sculptured decoration	Fruit decoration

1. ________________

2. ________________

3. ________________

4. ____________

5. ____________

6. ____________

2 Learn and say

A: Which type of cuisine decoration do you like?

B: I like ____________.

Let's learn

Section A

1 Listen to the dialogue and repeat 

外教有声

Lily: Bill, What are you doing?

Bill: I am trying to decorate my new designed dish.

Lily: Oh, great! Which type of decoration would you prefer?

Bill: I have no idea. Would you please give me some suggestions?

Lily: What are your ideas of making this dish?

Bill: A sweet-soured fruit and vegetable salad for mothers on Mother's Day, which means a sweet love rewarded to all mothers.

Lily: What a heart-warmed dish! How about using a tulip to decorate?

Bill: Good idea! Thank you very much!

2 Fill in the blanks with the verbs you see in the pictures

pour	sculpture	set	draw	pile up	sprinkle

1. ______ the strawberry

2. ______ the cake

3. ______ the pumpkin

4. ______ the pepper

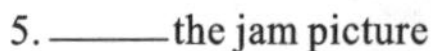

5. ______ the jam picture

6. ______ the jam

❸ **Put the sentences below into the right order in case you decorate a dish**

__________ Finally, try to make the dish a good presentation.

__________ And choose a right type of decoration for the dish.

__________ Then, dry the plate.

__________ First, choose a suitable plate and wash it.

__________ Next, pour the dish into the plate.

Section B

❶ **Read the short passage**

Nowadays people pay more attention to cuisine decoration. Because it can not only make the dish look delicious but also stir the desire of eating. When decorate a dish, you should pay attention to color, flavor, temperature, theme and idea of making the dish. If you want to be a qualified cook, you must work hard on your professional skill and learn to decorate dishes as well.

❷ **Answer the questions**

1. Why do people pay more attention to cuisine decoration?
2. What should you pay attention to when you decorate a dish?
3. If you want to be a qualified cook, what will you do?

扫码看答案

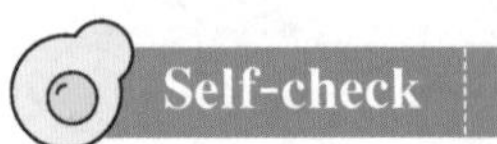

Words I have learned in this lesson are:

☐ cook ☐ cuisine ☐ dish ☐ draw
☐ flavor ☐ jam ☐ plate ☐ pile
☐ pour ☐ professional ☐ pumpkin ☐ qualified
☐ sculpture ☐ skill ☐ stir ☐ strawberry
☐ temperature ☐ theme ☐ turnip

I know __________ words.

Phrases and expressions I have learned in this lesson are:

☐ paste decoration ☐ plate decoration
☐ fruit decoration ☐ jam picture decoration
☐ sculptured decoration ☐ vegetable decoration

I know __________ phrases and expressions.

I can:

☐ identify types of cuisine decoration
☐ learn requirements of cuisine decoration
☐ make cuisine decoration

Unit 7

Service

Lesson 25 Service for Chinese Food

扫码看课件

Goal

You will be able to:

1. Know how to take an order in Chinese restaurant.
2. Master the serving order of Chinese food.

Warming up

1 Read and match

Bowl	Chopsticks	Plate
Table cloth	Chopstick holder	Spoon

1. ______

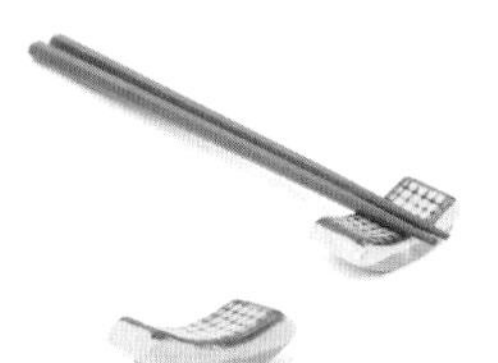

2. ______

3. ______

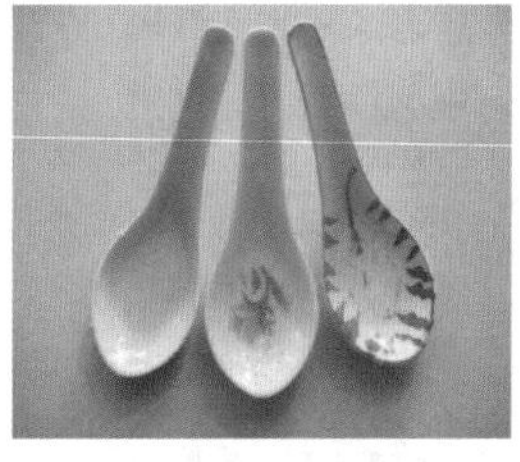

4. ______

5. ______

6. ______

2 Learn and repeat

A: Good morning, welcome to our restaurant. Do you have a reservation?

B: Yes.

A: May I know your name?

B: Bill.

A: I've found the reservation under the name of Bill. This way, please. Is this table fine?

Note

B: Fine, thanks.

A: Please wait a moment. I'll bring the menu to you soon.

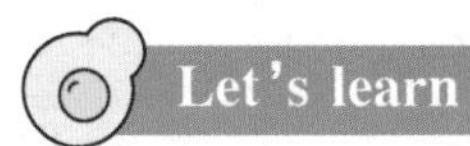

Let's learn

Section A

外教有声

❶ Listen to the dialogue and repeat 

Waiter: Excuse me, may I take your order now?

Guest: OK. But I know little about Chinese food. Can you tell me something about it?

Waiter: With pleasure, sir. Our restaurant serves Sichuan cuisine and Cantonese cuisine.

Guest: What's the difference between them?

Waiter: Cantonese cuisine is famous for its lightness, while Sichuan cuisine is very hot and spicy.

Guest: I'll try Cantonese cuisine, please. Can you recommend to me some specialties of your restaurant?

Waiter: Braised tomatoes with milk is our chef's recommendation. It's delicious and worth a try. Also, many guests have high comments on sliced boiled chicken in our restaurant.

Guest: OK. I'll take them.

Waiter: Would you like to drink something with your meal?

Guest: No, thanks.

Waiter: Please wait a moment, and I'll bring your dishes soon.

❷ Listen again and answer the follow questions

1. Is Cantonese cuisine hot and spicy?
2. What is Cantonese cuisine famous for?
3. What is the chef's recommendation?
4. Which dish do many guests have high comments on?

Section B

❶ Read the short passage

In Chinese restaurants, the serving order is unique: first cold dish, then hot dish; first stir-fried dish, then stewed dish. Fresh and light dishes are served before the sweet and spicy dishes. It takes time to prepare the Chinese food, so the guests are usually served with tea first. Then all the dishes

are served in succession: cold dish, hot dish, main course, soup, dessert and fruit.

2 Write down the appropriate words according to the pictures

1. ____________ 2. ____________ 3. ____________

4. ____________ 5. ____________ 6. ____________

3 Put the following pictures in the right order according to the serving order

1. ____________ 2. ____________ 3. ____________

4. ____________ 5. ____________ 6. ____________

扫码看答案

Self-check

Words I have learned in this lesson are:

- □ bowl
- □ plate
- □ chopsticks
- □ table cloth
- □ chopstick holder
- □ spoon
- □ lightness
- □ cold dish
- □ recommendation
- □ hot dish
- □ soup
- □ tea
- □ fruit
- □ dessert

I know __________ words.

Phrases and expressions I have learned in this lesson are:

- □ take one's order
- □ be famous for
- □ know little about
- □ worth a try

I know __________ phrases and expressions.

Note

I can:

□ know the serving order of Chinese food

Lesson 26 Service for Western Food

扫码看课件

Goal

You will be able to:

1. Master how to take an order in western restaurant.
2. Know the serving order of western food.

Warming up

❶ Read and match

Coffee cup	Dessert fork	Spoon
Fork	Wine glass	Knife

1. ____________ 2. ____________ 3. ____________

4. ____________ 5. ____________ 6. ____________

❷ Learn and say

A: Are you ready to order now?

B: Maybe you can recommend something for the main course.

A: Smoked salmon is great. I suggest you try that.

B: OK, I'll have one.

A: Good choice. What about an appetizer?

B: No, thanks.

A: OK. Would you like anything for dessert? We have apple pie and lemon pudding.

B: Lemon pudding, please.

A: So that's one smoked salmon and one lemon pudding. Is that right?

B: That's right. Thank you.

A: I'll bring them to you in a moment.

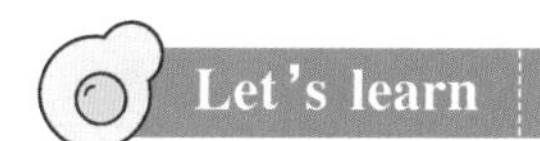

Let's learn

Section A

1 Listen to the dialogue and repeat 

外教有声

Waiter: This is your menu, sir.

Guest: Thanks.

Waiter: May I take your order now?

Guest: Yes. I'd like to have steak, potato wedges and creamy pumpkin soup.

Waiter: All right. How would you like your steak cooked?

Guest: Well done, please.

Waiter: Anything to drink, sir?

Guest: A glass of red wine, please.

Waiter: OK. Please wait a moment. And I'll be with you soon.

2 Decide true (T) or false (F)

(   ) 1. The guest ordered potato chips.

(   ) 2. The guest chose creamy pumpkin soup.

(   ) 3. The guest would like to have medium cooked steak.

Section B

1 Read the short passage

In western banquet, people often have seven course meals. They often start with appetizer like French baked snail. Appetizer, usually being salty and sour, is few in quantity and exquisite in quality. Then the soup is served. People usually have soup with a spoon because the soup is thicker than Chinese soup. Side dish is served after soup, including seafood like shell and fish. The side dish is followed by main course which usually comprises of meat and poultry. It can be beef, mutton, chicken and so on. Vegetables are served as salad after main course. Then people eat dessert such as

pudding, ice cream, cheesecake and so on. Finally, the meal would end up with the serving of tea or coffee.

❷ Decide true (T) or false (F)

(　　) 1. In western banquet, people often have six course meals.

(　　) 2. Soup is served after side dish.

(　　) 3. Side dish includes shell and fish.

❸ Put the following pictures in the right order according to the serving order

1. ______________

2. ______________

3. ______________

4. ______________

5. ______________

6. ______________

扫码看答案

Self-check

Words I have learned in this lesson are:

- ☐ steak
- ☐ fork
- ☐ spoon
- ☐ wine
- ☐ mashed potatoes
- ☐ salmon
- ☐ potato wedges
- ☐ creamy
- ☐ pumpkin
- ☐ well done
- ☐ appetizer
- ☐ snail
- ☐ quantity
- ☐ exquisite
- ☐ quality
- ☐ pudding
- ☐ ice cream
- ☐ cheesecake

I know __________ words.

Phrases and expressions I have learned in this lesson are:

- ☐ start with
- ☐ comprise of
- ☐ be served with
- ☐ such as
- ☐ end up with

I know __________ phrases and expressions.

I can:

- ☐ know the serving order of western food

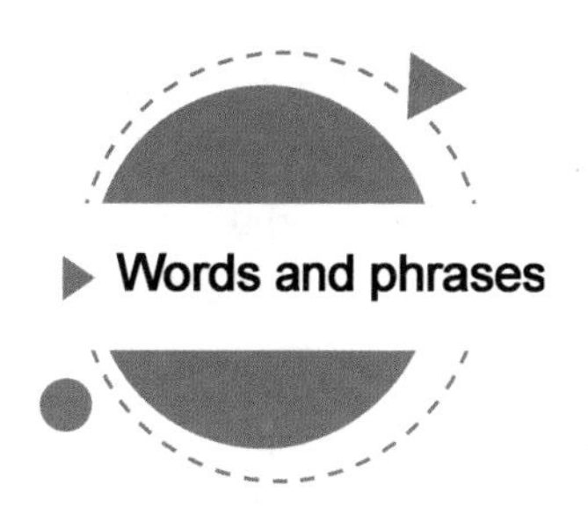

Words and phrases

Lesson 1

assistant [ə'sɪstənt] n. 助手，助理
adj. 助理的，辅助的，副的

chef [ʃef] n. 厨师，主厨

cook [kʊk] vt. 烹调，煮
vi. 烹调，做菜
n. 厨师，厨子

executive [ɪg'zekjʊtɪv] adj. 行政的，经营的，执行的

pastry ['peɪstrɪ] n. 油酥点心，面粉糕饼

roast [rəʊst] n. 烤肉，烘烤
v. 烤，烘焙，晒
adj. 烘烤的，烤过的

sauce [sɔːs] n. 酱汁，调味汁
vt. 给……调味，使……增加趣味

vegetable ['vedʒɪtəbl] n. 蔬菜，植物
adj. 蔬菜的，植物的

Lesson 2

automatic [ˌɔːtə'mætɪk] adj. 自动的
n. 自动装置

barbecue ['bɑːbɪkjuː] n. 烤肉
vt. 烧烤，烤肉

blender ['blendə(r)] n. 搅拌机

cabinet ['kæbɪnət] n. 橱柜，展览艺术品的小陈列室

chopsticks ['tʃɒpstɪks] n. 筷子

disinfection [ˌdɪsɪn'fekʃn] n. 消毒

dishwasher ['dɪʃˌwɒʃə(r)] n. 洗碗工，洗碟机

electronic [ˌɪlek'trɒnɪk] adj. 电子的

grill [grɪl] n. 烤架，铁格子，烧烤（食物）
vt. 拷问，（在烤架上）烤

grinder ['graɪndə(r)] n. [机]研磨机，研磨者，磨工，臼齿

microwave ['maɪkrəʊweɪv] n. 微波

oven [ˈʌvn]　n. 炉，灶；烤炉，烤箱
refrigerator [rɪˈfrɪdʒəreɪtə(r)]　n. 冰箱，冷藏库，制冰机，冷冻机
scale [skeɪl]　n. 刻度，量程，天平，磅秤
　v. 测量
slicer [ˈslaɪsə(r)]　n. 切片机
spoon [spuːn]　n. 匙，调羹，匙状物
　vt. 用匙舀起
steamer [ˈstiːmə(r)]　n. 轮船，蒸汽机，蒸笼
stove [stəʊv]　n. 火炉，窑，温室
　vt. 用火炉烤

Lesson 3

beet [biːt]　n. 甜菜，甜菜根，糖萝卜
broccoli [ˈbrɒkəlɪ]　n. 花椰菜，西蓝花
carrot [ˈkærət]　n. 胡萝卜
celery [ˈselərɪ]　n. 芹菜，香芹粉，芹菜籽
classify [ˈklæsɪfaɪ]　vt. 分类，归类
cucumber [ˈkjuːkʌmbə(r)]　n. 黄瓜，胡瓜
cut... into...　把……切成……
eggplant [ˈegplɑːnt]　n. 茄子
fruit [fruːt]　n. 成果，水果，果实
kiwifruit [ˈkiwiˌfrʊt]　n. 奇异果，猕猴桃
lettuce [ˈletɪs]　n. 莴苣，生菜
mango [ˈmæŋgəʊ]　n. 杧果，杧果树，泡菜
melon [ˈmelən]　n. 瓜（葫芦科甜瓜和西瓜数个品种果实的统称）
pea [piː]　n. 豌豆
peel [piːl]　vt. 剥皮，覆盖层脱落，剥落
　vi. 剥落，脱落，揭掉，表面起皮
　n. 果皮，蔬菜皮
pineapple [ˈpaɪnˌæpl]　n. 菠萝，凤梨
pour [pɔː(r)]　v. 不断流动，倒，倾泻
pour... into...　把……倒入……
pumpkin [ˈpʌmpkɪn]　n. 南瓜
strawberry [ˈstrɔːbərɪ]　n. 草莓，草莓色
string [strɪŋ]　n. 线，弦，细绳；一串，一行
turnip [ˈtɜːnɪp]　n. 芜菁，萝卜，芜菁作物

Lesson 4

be famous for　以……而出名
be worth of　值得
chicken breast　鸡胸
chicken thigh　鸡大腿

crisp [krɪsp] adj. 脆的
dice [daɪs] v. 将(食物等)切成丁
duck neck 鸭脖
flatten ['flæt(ə)n] vt. 使……平坦
fried peanuts 炸花生米
fry... until 将……炸至
fragrance ['freɪgrəns] n. 香味,芬芳
fry [fraɪ] v. 油炸,油煎
get rid of 处理掉
goose web 鹅掌
greasy ['gri:zɪ] adj. 油腻的,含脂肪多的
marinate ['mærɪneɪt] v. 腌制,腌渍
quail [kweɪl] n. 鹌鹑
guard [gɑ:d] vt. 保卫
put... into... 放入
recommend [ˌrekə'mend] vt. 推荐
Kung pao chicken 宫保鸡丁
starch [stɑ:tʃ] n. 淀粉
steam [sti:m] vi. 冒水汽,蒸
tender ['tendə(r)] adj. (食物)柔软的
gourd chicken 葫芦鸡
turn into 变成

Lesson 5

anise ['ænɪs] n. 大茴香
a spoon of 一勺
beef [bi:f] n. 牛肉
beef shank 牛腱
chewy ['tʃu:ɪ] adj. 有嚼劲的
cinnamon ['sɪnəmən] n. 肉桂皮
complex ['kɒmpleks] adj. 复杂的,合成的
cube [kju:b] v. 切成小方块
daily ['deɪlɪ] adj. 日常的,每日的
eggplant ['egplɑ:nt] n. 茄子
fierce [fɪəs] adj. 凶猛的,猛烈的
fragrance ['freɪgrəns] n. 香味,芬芳
hard [hɑ:d] adj. 努力的,坚硬的
lamb [læm] n. 羔羊,小羊,羊羔肉
pork [pɔ:k] n. 猪肉
red meat 红肉
refer to 参考,涉及,指的是
soft [sɒft] adj. 软的

Note

spare ribs　排骨
start with　以……开始
stir-fry [ˈstɜːfraɪ]　v. 翻炒，用旺火爆炒
streaky pork　五花肉
submerge [səbˈmɜːdʒ]　vt. 淹没，把……浸入
white meat　白肉
venison [ˈvenɪsən, ˈvenɪzən]　n. 鹿肉

Lesson 6

bass [beɪs]　n. 鲈鱼
be likely to　有……的可能性
catfish [ˈkætfɪʃ]　n. 鲇鱼
crucian [ˈkruːʃ(ə)n]　n. 鲫鱼
cut... open　切开
crab [kræb]　n. 螃蟹，蟹肉
dementia [dɪˈmenʃə]　n. 痴呆
eel [iːl]　n. 鳗鱼，鳝鱼
fatty acid　脂肪酸
fish [fɪʃ]　n. 鱼，鱼肉
fish line　鱼线
give up　放弃
grass carp　草鱼
hairtail [ˈheəteɪl]　n. 带鱼
intellectual development　智力发展
improve [ɪmˈpruːv]　vt. 改善，增进
last but not the least　最后但重要的
lifespan [ˈlaɪfspæn]　n. 寿命
nutritive value　营养价值
oyster [ˈɔɪstə(r)]　n. 牡蛎
promote [prəˈməʊt]　vt. 促进，提升
robust [rəʊˈbʌst]　adj. 强健的
scale [skeɪl]　vt. 刮鳞
　　n. 鳞
shrimp [ʃrɪmp]　n. 虾，小虾
turtle [ˈtɜːtl]　n. 龟，甲鱼
value [ˈvæljuː]　n. 值，价值
viscera [ˈvɪsərə]　n. 内脏
yellow croaker　黄鱼，黄花鱼

Lesson 7

acetic acid　醋酸
at once　立刻

absorb [əb'sɔːb] vt. 吸收,吸引
acid-base ['æsɪdbeɪs] n. 酸碱
balance ['bæləns] vt. 使平衡
bitter ['bɪtə(r)] adj. 苦的
calcium ['kælsɪəm] n. [化学]钙
chilli sauce 辣椒酱
condiment ['kɒndɪmənt] n. 调味品,佐料
digestion [daɪ'dʒestʃən] n. 消化
dry red chili 干红辣椒
easy to do 容易做
fatigue [fə'tiːg] n. 疲劳,疲乏
garlic ['gɑːlɪk] n. 大蒜
ginger ['dʒɪndʒə(r)] n. 姜
have a try 尝试
hot [hɒt] adj. 热的,辣的
kidney ['kɪdnɪ] n. 肾脏
lard oil 猪油
liver ['lɪvə(r)] n. 肝脏
mediate ['miːdɪət] vt. 调节
more than 超过
oyster oil 蚝油
preparation [ˌprepə'reɪʃ(ə)n] n. 准备
prickly ash 花椒
procedure [prə'siːdʒə(r)] n. 程序,手续
rapeseed oil 菜籽油
relieve [rɪ'liːv] vt. 解除,减轻
rice wine 黄酒,米酒
salt [sɔːlt] n. 盐
salty ['sɔːltɪ] adj. 咸的,含盐的
scallion ['skælɪən] n. 青葱
sour [saʊə(r)] adj. 酸的,发酵的
soy bean paste 豆瓣酱
soy sauce 酱油
stuff... into... 把……塞进……
sweet [swiːt] adj. 甜的
take in 吸收

Lesson 8

adjacent [ə'dʒeɪsənt] adj. 与……毗连的,邻近的
braised intestines in brown sauce 九转大肠
braised sea cucumber with scallion 葱烧海参
Confucius [kən'fjuːʃəs] n. 孔子

Confucianism [kən'fjuːʃənɪzəm] n. 儒教，孔子学说
date back 追溯
Dezhou braised chicken 德州扒鸡
steamed tofu stuffed with vegetables 一品豆腐
sweet and sour Yellow River carp 糖醋黄河鲤鱼

Lesson 9

aromatic [ˌærə'mætɪk] adj. 芳香的，有香味的
chili ['tʃɪlɪ] n. 辣椒
cube [kjuːb] n. 立方体
cuisine [kwi'ziːn] n. 烹饪，风味
delicious [dɪ'lɪʃəs] adj. 美味的，可口的
drain [dreɪn] v. 使……流走
flavor ['fleɪvə(r)] n. 味道，滋味，特色，风味
v. 给……调味，加味于
flour ['flaʊə(r)] n. 面粉
freckled ['frekəld] adj. 长有雀斑的，生有色斑的
garlic ['ɡɑːlɪk] n. 蒜，大蒜
ginger ['dʒɪndʒə(r)] n. 姜
in honor of 为了纪念
invention [ɪn'venʃən] n. 发明，创意，创造
literally ['lɪtərəlɪ] adv. 按字面，字面上
mince [mɪns] n. 绞碎的肉，肉馅
ordinary ['ɔːdɪnərɪ] adj. 普通的，平常的，平凡的
pass on 传递
pockmark ['pɒkmɑːk] n.（皮肤上的）麻点
pork [pɔːk] n. 猪肉
province ['prɒvɪns] n. 省
remove [rɪ'muːv] v. 移开，拿走
rise to fame 出名
scallion ['skæliən] n. 葱
shredded ['ʃredɪd] adj. 切成丝的
skimmer ['skɪmə(r)] n. 漏勺
slice [slaɪs] n.（切下食物的）薄片，片
specialty ['speʃəltɪ] n. 特色菜，特色食品
spicy ['spaɪsɪ] adj. 辣的，加有香料的
sprinkle ['sprɪŋkl] v. 撒，把……撒在……上
starch [stɑːtʃ] n. 淀粉
sticky ['stɪkɪ] adj. 黏稠的
stir [stɜː(r)] v. 搅动，搅拌
tasty ['teɪstɪ] adj. 美味的，可口的，好吃的
trace back to 追溯到

unique [ju'niːk] adj. 唯一的，独一无二的
wok [wɒk] n.（炒菜的）锅

Lesson 10

braise [breɪz] v. 煨
capital ['kæpɪtl] n. 首都，国都
center ['sentə(r)] n. 中心，中心区，中央
chili ['tʃɪlɪ] n. 辣椒
flavor ['fleɪvə] n. 味，滋味，特色，风味
fragrant ['freɪgrənt] adj. 香的，芳香的
medium ['miːdiəm] adj. 中等的，中号的
pepper ['pepə(r)] n. 胡椒粉，甜椒，柿子椒，灯笼椒
smoky ['sməʊkɪ] adj. 烟熏的，烟熏味的
stir [stɜː(r)] v. 搅动，搅和，搅拌
typical ['tɪpɪkl] adj. 典型的，有代表性的
on business 出差
spicy food 辣的食物
heard of 听说
stinky tofu 臭豆腐
steamed fish head with chopped pepper 剁椒鱼头
the center of ……的中心

Lesson 11

authentic [ɔː'θentɪk] adj. 真正的，真品的，真实的
dam [dæm] n. 水坝，拦河坝
greasy ['griːzɪ] adj. 多油的，油腻的，油性的，多脂的
local ['ləʊkl] adj. 地方的，当地的，本地的
mellow ['meləʊ] adj. 醇香的，甘美的
oily ['ɔɪlɪ] adj. 含油的，油腻的
official [ə'fɪʃəl] n. 要员，官员
poet ['pəʊɪt] n. 诗人
renovate ['renəveɪt] v. 修复，翻新，重新粉刷
shrimp [ʃrɪmp] n. 虾，小虾
shred ['ʃred] v. 切碎，撕碎
stuff [stʌf] v. 填满，装满，塞满，灌满，把……塞进（或填进）
tenderloin ['tendəlɔɪn] n.（牛、猪等的）里脊肉，嫩腰肉
typical ['tɪpɪkl] adj. 典型的，有代表性的，特有的
vinegar ['vɪnɪgə(r)] n. 醋

Lesson 12

boil [bɔɪl] v.（使）沸腾，煮沸，烧开
banquet ['bæŋkwɪt] n. 宴会，盛宴，筵席

dynasty [ˈdɪnəstɪ] n. 王朝，朝代
eel [iːl] n. 鳗，鳗鲡
emperor [ˈempərə(r)] n. 皇帝
material [məˈtɪərɪəl] n. 原料，(某一活动所需的)材料
monosodium glutamate 谷氨酸钠，味精，味素
nutrition [njuˈtrɪʃən] n. 营养，营养的补给
predecessor [ˈpriːdɪsesə(r)] n. 前任，原先的东西，被替代的事物
stew [stjuː] v. 炖，煨
seasoning [ˈsiːzənɪŋ] n. 调味品，佐料
scallion [ˈskæljən] n. 大葱
starch [stɑːtʃ] n. 淀粉
soy sauce 酱油
traditional [trəˈdɪʃənəl] adj. 传统的，习俗的

Lesson 13

Buddha jumping over the wall 佛跳墙
delicacies [ˈdelɪkəsɪz] n. 佳肴，美味(delicacy 的复数形式)
flavor [ˈfleɪvə(r)] n. 味，滋味，特色，风味
v. 给……调味
Fujian peanuts 闽生果
greasy [ˈgriːzɪ] adj. 多油的，油污的，油腻的，油性的，多脂的
litchi meat 荔枝肉
originated [əˈrɪdʒɪneɪtɪd] v. 起源，发源，创立(originate 的过去分词和过去式)
red wine fish steak 红糟鱼排
seven-star fish ball 七星鱼丸
white snow chicken 白雪鸡

Lesson 14

abalone [ˌæbəˈləʊnɪ] n. 鲍鱼
broth [brɒθ] n. (加入蔬菜的)肉汤，鱼汤
climate [ˈklaɪmət] n. 气候，气候区
mild [maɪld] adj. 温和的，和善的
moist [mɔɪst] adj. 微湿的，湿润的
recipe [ˈresɪpɪ] n. 烹饪法，食谱，方法
prosperous [ˈprɒspərəs] adj. 繁荣的，成功的，兴旺的
regional [ˈriːdʒənəl] adj. 地区的，区域的，地方的
relaxation [ˌriːlækˈseɪʃ(ə)n] n. 放松，休息，消遣
refer to 提到
roasted crispy suckling pig 脆皮乳猪
spring roll 春卷
boiled prawns 白灼虾
shrimp dumpling 虾饺

steamed rice roll 肠粉
sliced boiled chicken 白切鸡

Lesson 15

attention [ə'tenʃ(ə)n] n. 注意,注意力
braise [breɪz] vt. 炖,蒸
n. 文火炖熟的肉,赤鲷
crisp [krɪsp] adj. 脆的,干冷的,易碎的
dynasty ['dɪnəstɪ] n. 王朝,朝代
elegant ['elɪg(ə)nt] adj. 优雅的,雅致的
hotchpotch ['hɒtʃpɒtʃ] n. 杂烩,全家福,大杂烩
introduce [ɪntrə'djuːs] vt. 介绍
intact [ɪn'tækt] adj. 完好无缺的,原封不动的,未经触碰的
leek [liːk] n. 韭,[园艺]韭葱
mandarin fish 鳜鱼
originate [ə'rɪdʒɪneɪt] vt. 引起,创始,创作,开始,发生
vi. 起源于,来自,产生
rub [rʌb] vt. 擦,摩擦,用……擦,(使)相互摩擦
vi. 接触,摩擦,擦伤,在困境中持续下去
rub... into... 把……擀成……
remove [rɪ'muːv] vt. 移动,迁移,开除,调动
vi. 移动,迁移,搬家
section ['sekʃ(ə)n] n. 部分,部门,章节,区域
slice [slaɪs] n. 薄片,部分
smelly ['smelɪ] adj. 发出难闻气味的,有臭味的
smelly mandarin fish 臭鳜鱼
stewed soft shell turtle with ham 火腿炖甲鱼
temperature ['tempərətʃə(r)] n. 气温,体温,温度

Lesson 16

Japanese [ˌdʒæpə'niːz] adj. 日本(人)的,日语的
n. 日本人,日语
tempura ['tempuːrɑː] n. 天妇罗(日本菜肴)
sashimi ['sɑːʃɪmɪ] n. (日)生鱼片
sushi ['suːʃɪ] n. 寿司
ramen ['rɑːmen] n. 拉面,面条
fantastic [fæn'tæstɪk] adj. 奇异的,极好的,不可思议的
wheat [wiːt] n. 小麦,小麦色
flour ['flaʊə(r)] n. 面粉
vt. 撒粉于
alkaline ['ælkəlaɪn] adj. 碱性的,含碱的
instant ['ɪnst(ə)nt] adj. 立即的

Note

n. 瞬间
broth [brɒθ] n. 肉汤
gently ['dʒentlɪ] adv. 轻轻地，温柔地
ounce [aʊns] n. 盎司
package ['pækɪdʒ] n. 包，包裹
vt. 打包
silken ['sɪlk(ə)n] adj. 绸的，柔软的
dice [daɪs] v. 切成小方块
onion ['ʌnjən] n. 洋葱，洋葱头
sliced [slaɪst] adj. (食物)已切成薄片的
diagonally [daɪ'æg(ə)nəlɪ; daɪ'ægən(ə)lɪ] adv. 对角地，斜对地
inch [ɪntʃ] n. 英寸
piece [piːs] n. 块，件
teaspoon ['tiːspuːn] n. 茶匙，一茶匙的量
tablespoon ['teɪblspuːn] n. 大调羹，大汤匙，一餐匙的量
paste [peɪst] n. 面团，肉(或鱼等)酱(涂抹料或烹饪用)
medium ['miːdɪəm] adj. 中间的，半生熟的
saucepan ['sɔːspən] n. 炖锅，深平底锅
combine [kəm'baɪn] vt. 使联合，使结合
vi. 联合，结合
separate ['sepəreɪt; 'seprət] adj. 分开的，单独的，各自的
dashi ['dɑːʃiː] n. (日)鲣鱼汤
granule ['grænjuːl] n. 颗粒
layer ['leɪə] n. 层
vt. 把……分层堆放
vi. 形成或分成层次
simmer ['sɪmə] v. 炖，煨
n. 炖，即将沸腾的状态
whisk [wɪsk] v. 搅动，挥动
n. 搅拌器
miso soup 味噌汤
kobe steak 神户牛排
go out for 出去吃
come over 过来，顺便来访
fix for dinner 准备晚餐
be made with... 由……制成
make... different from others 使……与众不同
sliced... into... 切成……
bring to a boil 使沸腾
reduce heat to 减少热量，把火调小
stir in 一边加原料一边调和

add... to 加入，加到
simmer for 炖(多久)

Lesson 17

Korean [kə'riən] adj. 韩国人(语)的,朝鲜人(语)的
n. 韩国人(语),朝鲜人(语)
kimchi ['kɪmtʃɪ] n.(朝鲜语)朝鲜泡菜
grilled [grɪld] adj. 烤的
rib [rɪb] n. 肋骨,排骨
thick [θɪk] adj. 厚的
barbecue ['bɑːbɪkjuː] n. 烤肉
vt. 烧烤,烤肉
alongside [ə'lɒŋ'saɪd] prep. 在……旁边
adv. 在旁边
immediately [ɪ'miːdɪətlɪ] adv. 立刻
conj. 一……就……
strip [strɪp] n. 条状物
v. 剥去(外皮)
stockpot ['stɒkpɒt] n. 汤锅,杂烩锅
cover ['kʌvə(r)] vi. 覆盖
soak [səʊk] vt. 吸收
vi. 浸泡
refrigerated [rɪ'frɪdʒəreɪtɪd] adj. 冷冻的,冷却的
drain [dreɪn] v. 排水,流干
marinate ['mærɪneɪt] v. 腌制,腌渍,浸泡(食物)
blender ['blendə(r)] n. 搅拌机
puree ['pjʊəreɪ] n. 浓汤,果泥,菜泥
vt. 煮成浓汤或酱
sesame ['sesəmɪ] n. 芝麻
mixture ['mɪkstʃə(r)] n. 混合,混合物
overnight [ˌəʊvə'naɪt] n. 一夜的逗留
v. 过夜
tender ['tendə] adj.(食物)柔软的
crusty ['krʌstɪ] adj. 有壳的,像外壳一样的
garnish ['gɑːnɪʃ] v. 装饰(尤指食物)
boneless ['bəʊnləs] adj. 无骨的
skinless ['skɪnləs] adj. 无皮的
thigh [θaɪ] n. 大腿
shallot [ʃə'lɒt] n. 葱
chop [tʃɒp] v. 砍,斩碎
minced [mɪnst] adj. 切碎的,切成末的
mirin [mɪrɪn] n.(日本烹饪用的)米酒
chile ['tʃɪlɪ] n. 红番椒
syrup ['sɪrəp] n. 糖浆

Note

grated [ˈgreɪtɪd]　adj. 搓碎的
rinse [rɪns]　vt. 冲洗掉，漂净
　　n. 冲洗，漂洗
remove [rɪˈmuːv]　vt. 移动
　　vi. 移动
pat [pæt]　v.（用手掌）轻拍，轻拍……（使）成形
grill [grɪl]　n. 烤架，烧烤的菜肴（尤指烤肉）
　　v.（在烤架上）炙烤
brush [brʌʃ]　n. 刷子
　　v. 刷
surface [ˈsɜːfɪs]　n. 表面
　　adj. 表面的
crosswise [ˈkrɒswaɪz]　adv. 交叉地，斜地，成十字状地
grilled pork belly　烤五花肉
ginseng chicken soup　人参鸡汤
Korean scallion pancake　朝鲜葱饼
Korean barbecued chicken　韩式烤鸡肉
use... to do...　使用……去做……
cut... into...　把……切成……
cover with　覆盖
pull out　取出
use... for doing...　使用……做……
got it　明白了，知道了
per side　每面
pat dry　轻拍把水吸干
set aside　留置
make sure to do　确保
fire up　生火
turn golden brown　变成金黄色

Lesson 18

southeast [ˌsaʊθˈiːst]　adj. 东南的
　　n. 东南，东南方
Asian [ˈeɪʃən]　n. 亚洲人
　　adj. 亚洲的，亚洲人的
Singapore [ˌsɪŋəˈpɔː(r)]　n. 新加坡
curry [ˈkʌrɪ]　n. 咖喱食品，咖喱粉，咖喱
　　v. 用咖喱做菜
Hainanese chicken rice　海南鸡饭
satay [ˈsɑːteɪ]　n.（马来西亚、印度尼西亚的）加香烤肉
baste [beɪst]　v.（烹调时）浇卤汁于（肉等）上
commis [kɒmɪs]　n. 初级厨师，厨助，副手

cube [kjuːb] n. 立方,立方体
vt. 使成立方体
meanwhile ['miːnwaɪl] adv. 同时,其间
crush [krʌʃ] v. 压扁,压碎
clove [kləʊv] n. 丁香,小鳞茎
peel [piːl] v. 剥,削
n. 皮
heat [hiːt] n. 高温,压力
vt. 把……加热
skillet ['skɪlɪt] n. 煮锅,长柄平底煎锅
translucent [træns'ljuːs(ə)nt] adj. 透明的,半透明的
powder ['paʊdə(r)] n. 粉,粉末
vt. 使成粉末
coconut ['kəʊkənʌt] n. 椰子,椰子肉
rapidly ['ræpɪdlɪ] adv. 迅速地,很快地
reduce [rɪ'djuːs] vt. 减少,降低
vi. 减少,缩小
thicken ['θɪkən] vt. 使变厚,使……变复杂
vi. 变浓,变厚,变复杂
taste [teɪst] n. 味道
vt. 尝
vi. 尝起来,有……的味道
coriander [ˌkɒrɪ'ændə(r)] n. 芫荽,香菜
marinade [ˌmærɪ'neɪd] n. 腌泡汁
vt. 腌泡
bamboo [ˌbæm'buː] n. 竹,竹子
vt. 为……装上篾条
adj. 竹制的
skewer ['skjuːə(r)] n. 烤肉叉子
vt. 串住
basting ['beɪstɪŋ] n. (烤肉时的)涂油脂
v. 涂以油脂
stalk [stɔːk] n. (植物的)茎,秆
lemon grass 柠檬香草,柠檬草,香茅
turmeric ['tɜːmərɪk] n. 姜黄,姜黄根,姜黄根粉
chili ['tʃɪlɪ] n. 红辣椒,辣椒
blend [blend] v. 混合
processor ['prəʊsesə(r)] n. 加工者
food processor 食物料理机
fridge [frɪdʒ] n. 电冰箱
thread [θred] n. 线
vt. 穿线于
vi. 穿透过

Note

Bak Kut Teh　肉骨茶
Tom yum　冬阴功汤
set . . . aside for　放置一边
show. . . how to. . .　展示如何操作
heat. . . in. . .　用……工具加热
over medium-high heat　中高温
changes from. . . to. . .　从……变成……
simmering rapidly　迅速升温
each side　每边,每面
both sides　两边,两面

Lesson 19

authentic [ɔːˈθentɪk]　adj. 真的,真正的,可信的,可靠的,有根据的
Cappadocia [ˌkæpəˈdəʊʃə]　n. [地名][土耳其]卡帕多西亚
crock [krɒk]　n. 坛子,瓦罐
Cappadocia crock beef　土耳其瓦罐牛肉
dolma [ˈdɒlmə]　n. 朵尔玛(土耳其馅饼)
dumpling [ˈdʌmplɪŋ]　n. 饺子,汤团,面团
dough [dəʊ]　n. 生面团
flour [ˈflaʊə(r)]　n. 面粉,粉状物质
all-purpose flour　中筋面粉
kebab [kəˈbɑːb]　n. (印度)烤肉串,烤腌羊肉串
mayonnaise [ˌmeɪəˈneɪz]　n. 蛋黄酱
mince [mɪns]　vt. 切碎,剁碎
paprika [ˈpæprɪkə]　n. 辣椒粉,红辣椒
parsley [ˈpɑːslɪ]　n. 欧芹,荷兰芹
pinch [pɪntʃ]　vt. 捏,掐
rotate [rəʊˈteɪt]　v. 轮流,(使)旋转
snack [snæk]　n. 小吃,快餐
sprinkle [ˈsprɪŋkl]　v. 撒,用……点缀
Turkey [ˈtɜːkɪ]　n. 土耳其
Turkish [ˈtɜːkɪʃ]　n. 土耳其语,土耳其人
　adj. 土耳其的,土耳其人的,土耳其语的
Turkish pide　土耳其比萨船
topping [ˈtɒpɪŋ]　n. 糕点上的装饰配料
vertical [ˈvɜːtɪk(ə)l]　adj. 垂直的,直立的
wrap [ræp]　v. 包,裹
yeast [jiːst]　n. 酵母,发酵剂

Lesson 20

western [ˈwestən]　adj. 西方的
　n. 西方人

pasta ['pæstə] n. 意大利面食，面团
sandwich ['sænwɪdʒ] n. 三明治
caviar ['kævɪɑːr] n. 鱼子酱
grate [greɪt] v. 磨碎(食物)
French [frentʃ] adj. 法国的，法国人的
n. 法国人
chop [tʃɒp] v. 剁碎，砍
n. (猪、羊等)排骨
sauté ['səʊteɪ] v. 嫩煎，炒
cheese [tʃiːz] n. 奶酪，干酪
slice [slaɪs] n. 薄片，菜刀
vt. 切下，将……切成薄片
vi. 切开
stir [stɜː(r)] v. 搅拌
constantly ['kɒnstəntlɪ] adv. 不断地，时常地
bay [beɪ] n. 月桂
thyme [taɪm] n. 百里香
flavor ['fleɪvə(r)] n. 风味，香料，滋味
vt. 加味于
pour [pɔː] v. 灌，倒
sprinkle ['sprɪŋkl] v. 撒，用……点缀
salamander ['sæləˌmændə(r)] n. 烤箱
foie gras 鹅肝酱
beef tartare 鞑靼牛肉
roast turkey 烤火鸡
pan fry 用平底锅(以少量油)煎(或炒)
use for 用于
how to make 如何制作
can't wait to 迫不及待地做
main ingredients for... are... 主要材料是
heat... to a boil 把……加热至沸腾
lower the heat 降低热量，将火调小
in the meantime 在此期间
pour... into... 倒入
top... with... 把(某物)放在……的上面
put... in... 放入
mix well 调匀，搅匀
garnish the plate with... 用……装盘

Lesson 21

bun [bʌn] n. 小圆面包
be popular with 受……欢迎

Note

chef [ʃef]　n. 厨师，大师傅
Chinese pastries　中式面点
dough [dəʊ]　n. 生面团
dumpling [ˈdʌmplɪŋ]　n. 汤团，面团，饺子
dim sum　(汉)点心
float [fləʊt]　v. 使漂浮，浮动，漂流，飘动，飘移
fry... till　油炸……直到
hot to serve　趁热吃
knead [niːd]　vt. 揉，捏，捏制
knead... into...　把……捏成……
lunar new year　农历新年
marvellous [ˈmɑːv(ə)ləs]　adj. 不可思议的，惊人的
melt [melt]　vt. 使融化，使熔化，使软化
mill [mɪl]　v. 碾磨，磨细
mould [məʊld]　v.（用模具）浇铸，塑造
pastry [peɪstrɪ]　n. 油酥点心，面粉糕饼
pudding [ˈpʊdɪŋ]　n. 布丁
reunion [riːˈjuːnjən]　n. 重聚
remove [rɪˈmuːv]　vt. 移动，迁移
sesame [ˈsesəmɪ]　n. 芝麻
Spring festival　春节
steep [stiːp]　vt. 泡，浸
steam [stiːm]　vi. 蒸
symbolize [ˈsɪmbəlaɪz]　vt. 象征

Lesson 22

add [æd]　vt. 增加，添加
add... into...　把……添加入……
bake... till...　烘烤……直到……
biscuit [ˈbɪskɪt]　n. 饼干，小面包
blow out　吹熄
can't wait to　等不及做
celebrate [ˈselɪbreɪt]　vt. 庆祝，举行
celebrate... with...　用……来庆祝……
container [kənˈteɪnə(r)]　n. 容器
dessert [dɪˈzɜːt]　n. 甜点，甜品
doughnut [ˈdəʊnʌt]　n. 油炸圈饼
egg yolk　蛋黄
extra [ˈekstrə]　adj. 额外的
fragrant [ˈfreɪgrənt]　adj. 芳香的
green tea powder　抹茶粉
have a try　试一试

Note

it is said that　据说
ingredient [ɪn'griːdjənt]　n. 材料，佐料
kid [kɪd]　n. 小孩
low gluten flour　低筋粉
macaroon [ˌmækə'ruːn]　n. 蛋白杏仁饼干，马卡龙
oven ['ʌvn]　n. 炉，灶；烤炉，烤箱
patient ['peɪʃ(ə)nt]　adj. 有耐心的
pastry bag　裱花袋
pie [paɪ]　n. 馅饼
preheat [priː'hiːt]　vt. 预先加热
prepare... for...　为……准备……
put... into...　把……放进……
recommendation [ˌrekəmen'deɪʃ(ə)n]　n. 推荐，建议
romantic [rəʊ'mæntɪk]　adj. 浪漫的，多情的，空想的
sift [sɪft]　vt. 筛，过滤
smooth [smuːð]　adj. 顺利的，光滑的，平稳的
squeeze [skwiːz]　v. 挤，压榨
stir [stɜː(r)]　v. 搅拌
stir to blend　搅拌至光滑
suggestion [sə'dʒestʃən]　n. 建议
tart [tɑːt]　n. 蛋挞
puff pastry　千层饼，松饼
tiramisu [ˌtɪrə'miːsuː]　n. 提拉米苏
turn to　变成

Lesson 23

chestnut ['tʃesnʌt]　n. 栗子，栗色，栗树，栗色马
adj. 栗色的
clam [klæm]　n. 蚌，蛤
liver ['lɪvə(r)]　n. 肝脏
mutton ['mʌt(ə)n]　n. 羊肉
pour [pɔː(r)]　v. 倒，倾泻，蜂拥而来
n. 倾泻
sautéed [səʊ'teɪd]　adj. 嫩煎的
scallion ['skælɪən]　n. 小洋葱，冬葱
slice [slaɪs]　vt. 切下，把……分成部分，将……切成薄片
n. 薄片
spinach ['spɪnɪdʒ]　n. 菠菜
stewed [stjuːd]　adj. 炖的，烂醉的
steamed [stiːmd]　adj. 蒸熟的

Note

Lesson 24

choose [tʃuːz] vt. 挑选
vi. 选择，进行挑选

decoration [ˌdekəˈreɪʃ(ə)n] n. 装饰，装潢，装饰品，装饰图案，装饰风格，奖章

delicious [dɪˈlɪʃəs] adj. 美味的，可口的

jam [dʒæm] n. 果酱

paste [peɪst] n. 面团，糊状物，糊
vt. 裱糊，粘贴

pepper [ˈpepə(r)] n. 胡椒粉，辣椒

picture [ˈpɪktʃə(r)] vt. 使图示化，用图画说明(或表示)

plate [pleɪt] n. 盘子，碟子

pumpkin [ˈpʌmpkɪn] n. 南瓜，南瓜的果肉

sculptured [ˈskʌlptʃəd] adj. 具刻纹的，用刻纹装饰的

strawberry [ˈstrɔːbərɪ] n. 草莓，草莓色

Lesson 25

bowl [bəʊl] n. 碗，钵

chopsticks [ˈtʃɒpstɪks] n. 筷子

chopstick holder n. 筷架

comment [ˈkɒment] n. 评论

menu [ˈmenjuː] n. 菜单

plate [pleɪt] n. 盘子，碟子

recommend [ˌrekəˈmend] v. 推荐

reservation [ˌrezəˈveɪʃn] n. 预订

specialty [ˈspeʃəltɪ] n. 特色食品，特产

spicy [ˈspaɪsɪ] adj. 辛辣的，刺激的

spoon [spuːn] n. 勺，匙

stewed [stjuːd] adj. 炖的，煨的

succession [səkˈseʃən] n. 一连串，一系列，连续的(人或事物)

table cloth n. 桌布

Lesson 26

appetizer [ˈæpɪtaɪzə(r)] n. 开胃菜

banquet [ˈbæŋkwɪt] n. 宴会，盛宴

comprise [kəmˈpraɪz] v. 包括，包含，由……组成

creamy [ˈkriːmɪ] adj. 像奶油的，含乳脂的

dessert [dɪˈzɜːt] n. 甜点，甜品

exquisite [ɪkˈskwɪzɪt] adj. 精美的，精致的

mashed potatoes n. 土豆泥

potato wedges n. 土豆角(一种零食)
poultry [ˈpəʊltrɪ] n. 家禽
pudding [ˈpʊdɪŋ] n. 布丁
pumpkin [ˈpʌmpkɪn] n. 南瓜
quality [ˈkwɒlətɪ] n. 质量,品质
quantity [ˈkwɒntətɪ] n. 数量,数目
recommend [rekəˈmend] v. 推荐
salad [ˈsæləd] n. 沙拉
salmon [ˈsæmən] n. 三文鱼
snail [sneɪl] n. 蜗牛
steak [steɪk] n. 牛排

Note

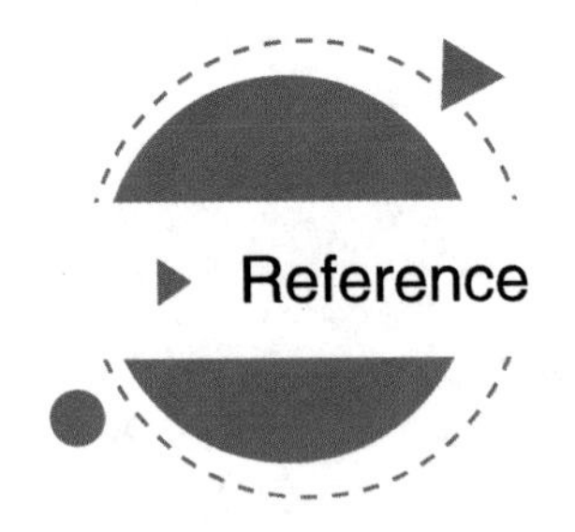

[1] 姜玲.厨师岗位英语[M].北京:旅游教育出版社,2008.

[2] 吴蓓蓓.餐饮服务行业实用英语对话及词汇手册[M].北京:中国水利水电出版社,2009.

[3] 赵丽.烹饪英语[M].北京:北京大学出版社,2010.

[4] 李柏红,张小玲.烹饪专业英语[M].北京:中国商业出版社,2006.

[5] 杜纲.烹饪英语[M].重庆:重庆大学出版社,2013.